Work with ionising radiation

Ionising Radiations Regulations 1999

APPROVED CODE
OF PRACTICE
AND GUIDANCE

L121

HSE BOOKS

First published 2000

ISBN 0 7176 1746 7

Approved Code of Practice and guidance

This Code has been approved by the Health and Safety Commission, with the consent of the Secretary of State. It gives practical advice on how to comply with the law. If you follow the advice you will be doing enough to comply with the law in respect of those specific matters on which the Code gives advice. You may use alternative methods to those set out in the Code in order to comply with the law.

However, the Code has a special legal status. If you are prosecuted for breach of health and safety law, and it is proved that you did not follow the relevant provisions of the Code, you will need to show that you have complied with the law in some other way or a court will find you at fault.

The Regulations and Approved Code of Practice (ACOP) are accompanied by guidance which does not form part of the ACOP. Following the guidance is not compulsory and you are free to take other action. But if you do follow the guidance you will normally be doing enough to comply with the law. Health and safety inspectors seek to secure compliance with the law and may refer to this guidance as illustrating good practice.

For convenience, the text of the Regulations is set out in *italic* type, with the ACOP in **bold** type and the accompanying guidance in normal type.

Contents

General

The Ionising Radiations Regulations 1999 (IRR99),[1] made under the Health and Safety at Work etc Act 1974 (HSW Act),[2] implement the majority of the Basic Safety Standards Directive 96/29/Euratom[3] (BSS Directive) in Great Britain (Northern Ireland publishes separate regulations). From 1 January 2000, they replace the Ionising Radiations Regulations 1985 (IRR85) (except for regulation 26 (special hazard assessments))which were made in response to the 1980 BSS Directive 80/836/Euratom (as amended by 84/467/Euratom). The main aim of the Regulations and the supporting Approved Code of Practice (ACOP) is to establish a framework for ensuring that exposure to ionising radiation arising from work activities, whether from man-made or natural radiation and from external radiation (eg X-ray set) or internal radiation (eg inhalation of a radioactive substance), is kept as low as reasonably practicable and does not exceed dose limits specified for individuals.

IRR99 also:

(a) replace the Ionising Radiations (Outside Workers) Regulations 1993 (OWR93), which were made to implement the Outside Workers Directive 90/641/Euratom; and

(b) implement a part of the Medical Exposures Directive 97/43/Euratom[4] in relation to equipment used in connection with medical exposures (the Health Departments in England, Wales, Scotland and Northern Ireland have responsibility for implementing the majority of that Directive).

The guidance which accompanies the Regulations and ACOP in this publication gives detailed advice about the scope and duties of the requirements imposed by IRR99. It is aimed at employers with duties under the Regulations but should also be useful to others such as radiation protection advisers, health and safety officers, radiation protection supervisors and safety representatives.

The Basic Safety Standards Directive

The 1996 Basic Safety Standards Directive (BSS Directive) reflects the 1990 recommendations of the International Commission on Radiological Protection (ICRP60). The Directive lays down basic safety standards for the protection of the health of workers and the general public against the dangers arising from ionising radiation. Implementation of the BSS Directive in Great Britain is achieved by a mixture of revised regulation (eg IRR99), existing legal provisions, such as the Nuclear Installations Act 1965[5] (NIA65) and the Radioactive Substances Act 1993[6] (RSA93), and new provisions, for example proposed regulations on emergency preparedness.

Origins of IRR99

Although the Ionising Radiations Regulations 1985 (IRR85) were introduced to implement most of the provisions of the 1980 BSS Directive (as amended in 1984), there was also a need to consolidate some of the provisions of two sets of regulations made under the Factories Act 1961 which predated the introduction of the Health and Safety at Work etc Act 1974. Those Regulations were the Ionising Radiations (Unsealed Radioactive Substances) Regulations 1968 and the Ionising Radiations (Sealed Sources) Regulations 1969. As a consequence, certain provisions of IRR85 - and therefore of IRR99 - have their origins in domestic legislation, rather than European directives.

Besides the need to implement the 1996 BSS Directive,[3] key influences on the development of IRR99 can be summarised as:

(a)   retaining the many effective features of IRR85 and approaching revision in an evolutionary way;

(b)   reviewing those IRR85 provisions of domestic rather than European origin to consider which should be retained;

(c)   improving the structure and transparency of IRR85;

(d)   reviewing a decade of working with IRR85 and, where appropriate, simplifying and clarifying amendments for particular provisions; and

(e)   following Better Regulation principles, ie ensuring that the Regulations are necessary, fair, effective, balanced and enjoy a broad degree of public confidence.

Figure 1 represents many of the regulatory developments that led to IRR99.

In the light of the 1996 BSS Directive and these considerations, IRR99:

(a)   introduce prior authorisation for certain uses of X-ray sets and accelerators from 13 May 2000 (regulation 5 of IRR99);

(b)   build on the requirement for risk assessment under the Management of Health and Safety at Work Regulations 1999[7] (regulation 7 of IRR99);

(c)   enhance the requirement to keep exposures as low as reasonably practicable (regulations 8 to 10 of IRR99);

(d)   specify lower dose limits for employees and members of the public (regulation 11 of IRR99);

(e)   bring in a system for explicit recognition of radiation protection advisers (RPAs) and specify circumstances where suitable RPAs should be consulted (regulations 2(1) and 13 of IRR99);

(f)   give additional flexibility on the designation of controlled and supervised areas and introduce modified requirements for those areas (regulations 16 to 18 of IRR99); and

(g)   modify the requirements for designation of classified persons, for assessing and recording the doses they receive and for medical surveillance.

They do so in a revised framework designed to help bring out the radiation protection concepts. The Regulations also provide various transitional arrangements.

Scope of the revised Regulations

The scope of application of IRR99[1] is set out in regulation 3. The Regulations apply to three categories of work - practices, work in radon atmospheres at concentrations above a specified action level, and work with materials containing naturally occurring radionuclides. The definition of 'practice' (regulation 2(1)) mainly covers work with artificial sources including both man-made radioactive substances and the operation of certain electrical equipment, such as X-ray sets, which emit ionising radiations. It also includes

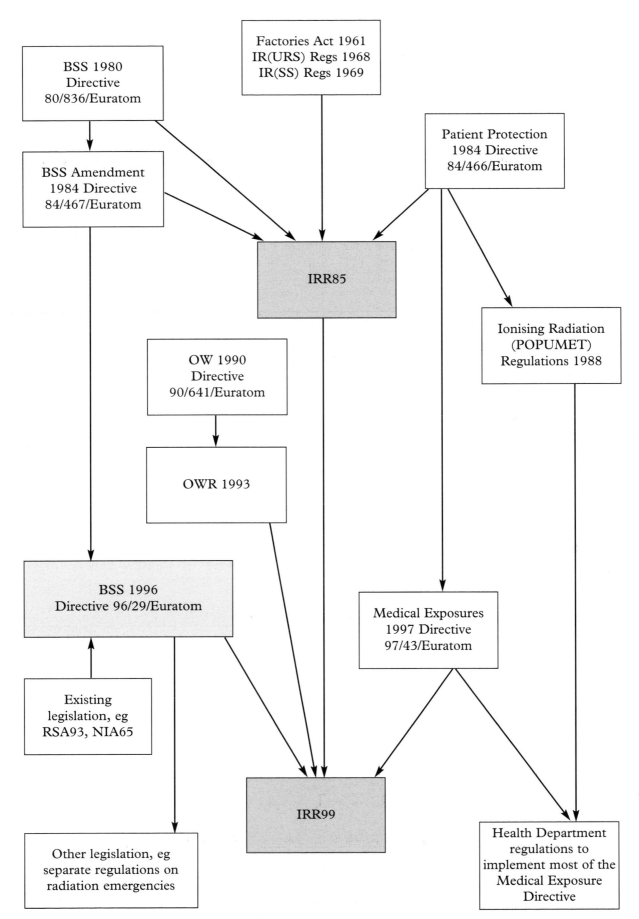

Figure 1 Regulatory developments relevant to IRR99

work with materials containing naturally occurring radionuclides but only where they are being processed for their 'radioactive, fissile or fertile properties'. This means that all activities in the nuclear fuel cycle are subject to the full control regime for practices.

The Regulations also apply to:

(a) work carried out in mines and other work where the construction of the workplace, or the ventilation provided, is insufficient to keep the concentration of radon-222 gas below levels specified in regulation 3(1)(b); and

(b) some forms of work with naturally occurring radioactive material with the potential to cause significant exposure.

Under regulation 3, certain provisions of the Regulations do not apply to people undergoing medical exposures; separate Regulations to implement most of the Medical Exposures Directive will apply in these situations.

Principal duty holders

The main duty holders are the same as in IRR85, although the term 'radiation employer' is used in place of 'employer who undertakes work with ionising radiation'. Radiation employers are the main subset of the wider group of employers who have duties under these Regulations (see Table 1). They are the employers who have specific duties to restrict the exposure of others because they are undertaking work with ionising radiation (and thereby creating possible risks which they are in the best position to control).

Certain regulations, for example regulation 18, place duties on the employer who designates an area as a controlled or supervised area. On most occasions this employer will also be a radiation employer, but the essence of the duty lies in the ability to control the area. The duties in regulation 32 rest with the 'employer who has to any extent control of equipment or apparatus used in connection with a medical exposure'. Again, this will usually be a radiation employer but the key aspect of the duty is that the particular employer is in a position to control the use of the equipment or apparatus.

Table 1 Duties on 'employers' under IRR99

Duty holder	Relevant regulations
Any employer	8(7), 11, 14, 15, 20 to 24 and 26
Radiation employer	5 to 7, 8(1) to (6), 9 to 10, 12, 13, 17, 25, 27 to 30, 32(6 to 7)
Employer in control of an area/who designates an area	16, 18 and 19
Employer in control of equipment	32(1) to (5)
Manufacturer or supplier of articles etc	31(1)
Installer	31(2)

Holders of a nuclear site licence under the Nuclear Installations Act 1965[5] are in a special position. Duties under these Regulations imposed on the employer are also imposed on the licensee to the extent that those duties relate to the licensed site (regulation 4(3) IRR99).

Duties of employees

Regulation 2(2) extends interpretation of the term 'employee' in the same way as IRR85, beyond its immediate meaning, to self-employed people and trainees. All employee duties have been grouped together in regulation 34 to allow employees to be clearer about their own duties under these Regulations.

Duties of self-employed people

A self-employed person who works with ionising radiation will simultaneously have certain duties under these Regulations both as an employer and as an employee. For example, self-employed persons may need to take such steps as:

(a) carrying out assessments under regulation 7;

(b) providing control measures under regulation 8 to restrict exposure;

(c) designating themselves as classified persons under regulation 20;

(d) making suitable arrangements under regulation 21 with one or more approved dosimetry services (ADS) for assessment and recording of doses they receive;

(e) obtaining a radiation passbook and keeping it up to date in accordance with regulation 21, if they carry out services as an outside worker;

(f) making arrangements for their own training as required by regulation 14; and

(g) ensuring they use properly any dosemeters provided by an ADS as required by regulation 34.

Apparently self-employed people

Although only the courts can give an authoritative interpretation of law, in considering the application of these regulations and guidance to people working under another's direction, the following should be considered.

If people working under the control and direction of others are treated as self-employed for tax and national insurance purposes they may nevertheless be treated as their employees for health and safety purposes. It may therefore be necessary to take appropriate action to protect them. If any doubt exists about who is responsible for the health and safety of a worker this could be clarified and included in the terms of a contract. However, remember, a legal duty under section 3 of the Health and Safety at Work etc Act 1974 (HSW Act)[2] cannot be passed on by means of a contract and there will still be duties towards others under section 3 of HSW Act. If such workers are employed on the basis that they are responsible for their own health and safety, legal advice should be sought before doing so.

Outside workers

Outside workers are employees designated as classified persons who carry out services in a controlled area designated by another employer. As the provisions relating to outside workers are only an elaboration of the general radiation protection regime, the requirements in IRR99 specific to this subset of employees appear in the appropriate regulation, rather than in a separate section for outside workers. As an example, the duty to provide a radiation passbook is included in the list of arrangements that employers must make with their ADS for record-keeping (regulation 21(5)).

A major change is that an outside worker's employer will no longer have to provide a new radiation passbook if that outside worker already has a valid one issued by a previous employer. The new employer will be able to take over the existing passbook, as long as they enter their own details under 'employer'.

Regulation 1

Regulation

1

Citation and commencement

These Regulations may be cited as the Ionising Radiations Regulations 1999 and shall come into force -

  (a)   as respect all regulations except for regulation 5, on 1st January 2000; and

  (b)   as respects regulation 5, on the 13th May 2000.

Regulation 2

Regulation

2(1)

Interpretation

  (1)   In these Regulations, unless the context otherwise requires -

"accelerator" means an apparatus or installation in which particles are accelerated and which emits ionising radiation with an energy higher than 1MeV;

"appointed doctor" means, subject to regulation 39(5) (which relates to transitional provisions), a registered medical practitioner who is for the time being appointed in writing by the Executive for the purposes of these Regulations;

"approved" means approved for the time being in writing for the purposes of these Regulations by the Health and Safety Commission or the Executive, as the case may be, and published in such form as the Health and Safety Commission or the Executive respectively considers appropriate;

"approved dosimetry service" means, subject to regulation 39(3) (which relates to transitional provisions), a dosimetry service approved in accordance with regulation 35;

"calendar year" means a period of 12 calendar months beginning with the 1st January;

"classified person" means -

  (a)   a person designated as such pursuant to regulation 20(1); and

  (b)   in the case of an outside worker employed by an undertaking in Northern Ireland or in another member State, a person who has been designated as a Category A exposed worker within the meaning of Article 21 of the Directive;

"comforter and carer" means an individual who (other than as part of his occupation) knowingly and willingly incurs an exposure to ionising radiation resulting from the support and comfort of another person who is undergoing or who has undergone any medical exposure;

"contamination" means the contamination by any radioactive substance of any surface (including any surface of the body or clothing) or any part of absorbent objects or materials or the contamination of liquids or gases by any radioactive substance;

"controlled area" means -

  (a)   in the case of an area situated in Great Britain, an area which has been so designated in accordance with regulation 16(1); and

7

(b) in the case of an area situated in Northern Ireland or in another member State, an area subject to special rules for the purposes of protection against ionising radiation and to which access is controlled as specified in Article 19 of the Directive;

"the Directive" means Council Directive 96/29/Euratom[a] laying down basic safety standards for the protection of the health of workers and the general public against the dangers arising from ionising radiation;

"dose" means, in relation to ionising radiation, any dose quantity or sum of dose quantities mentioned in Schedule 4;

"dose assessment" means the dose assessment made and recorded by an approved dosimetry service in accordance with regulation 21;

"dose constraint" means a restriction on the prospective doses to individuals which may result from a defined source;

"dose limit" means, in relation to persons of a specified class, the limit on effective dose or equivalent dose specified in Schedule 4 in relation to a person of that class;

"dose rate" means, in relation to a place, the rate at which a person or part of a person would receive a dose of ionising radiation from external radiation if he were at that place being a dose rate at that place averaged over one minute;

"dose record" means, in relation to a person, the record of the doses received by that person as a result of his exposure to ionising radiation, being the record made and maintained on behalf of the employer by the approved dosimetry service in accordance with regulation 21;

"employment medical adviser" means an employment medical adviser appointed under section 56 of the Health and Safety at Work etc. Act 1974;

"the Executive" means the Health and Safety Executive;

"external radiation" means, in relation to a person, ionising radiation coming from outside the body of that person;

"health record" means, subject to regulation 39(7) (which relates to transitional provisions), in relation to an employee, the record of medical surveillance of that employee maintained by the employer in accordance with regulation 24(3);

"internal radiation" means, in relation to a person, ionising radiation coming from inside the body of that person;

"ionising radiation" means the transfer of energy in the form of particles or electromagnetic waves of a wavelength of 100 nanometres or less or a frequency of 3×10^{15} hertz or more capable of producing ions directly or indirectly;

"licensee" has the meaning assigned to it by section 26(1) of the Nuclear Installations Act 1965[b];

"local rules" means rules made in accordance with regulation 17;

(a) OJ No L159, 29.6.96 p.l.
(b) 1965 c.57; relevant amending instruments are SI 1974/2056 and SI 1990/1918.

"maintained", where the reference is to maintaining plant, apparatus, equipment or facilities, means maintained in an efficient state, in efficient working order and good repair;

"medical exposure" means exposure of a person to ionising radiation for the purpose of his medical or dental examination or treatment which is conducted under the direction of a suitably qualified person and includes any such examination for legal purposes and any such examination or treatment conducted for the purposes of research;

"member State" means a member State of the Communities;

"outside worker" means a classified person who carries out services in the controlled area of any employer (other than the controlled area of his own employer);

1   An outside worker is a classified person who is carrying out services in a controlled area but who does not have an individual contract of employment with the employer who has responsibility for that area (as distinct from any contract for service between their own employer and the employer responsible for the area). 'Services' is an undefined term, but implies benefit to the employer responsible for the controlled area. Classified persons who are based at one site but who visit other sites which are all part of their employer's organisation are not outside workers. However, where each site is controlled by a different subsidiary company (ie a separate legal entity) then the classified persons will be outside workers on sites other than their base.

2   Self-employed contractors will have responsibilities both as outside workers and as employers when they perform services in the controlled areas of other employers (see regulation 2(2)).

"overexposure" means any exposure of a person to ionising radiation to the extent that the dose received by that person causes a dose limit relevant to that person to be exceeded or, in relation to regulation 26(2), causes a proportion of a dose limit relevant to any employee to be exceeded;

"practice" means work involving -

> *(a)   the production, processing, handling, use, holding, storage, transport or disposal of radioactive substances; or*

> *(b)   the operation of any electrical equipment emitting ionising radiation and containing components operating at a potential difference of more than 5kV,*

which can increase the exposure of individuals to radiation from an artificial source, or from a radioactive substance containing naturally occurring radionuclides which are processed for their radioactive, fissile or fertile properties;

"radiation accident" means an accident where immediate action would be required to prevent or reduce the exposure to ionising radiation of employees or any other persons;

"radiation employer" means an employer who in the course of a trade, business or other undertaking carries out work with ionising radiation and, for the purposes of regulations 5, 6 and 7, includes an employer who intends to carry out such work;

3   Radiation employers are essentially those employers who work with ionising radiation, ie they carry out:

(a)   a practice (see definition in regulation 2(1)); or

(b)   work in places where the radon gas concentration exceeds the values in regulation 3(1)(b); or

(c)   work with radioactive substances containing naturally occurring radionuclides not covered by the definition of a practice (see paragraph 11).

"radiation passbook" means -

    (a)   in the case of an outside worker employed by an employer in Great Britain -

        (i)   a passbook approved by the Executive for the purpose of these Regulations; or

        (ii)   a passbook to which regulation 39(4) (transitional provisions) applies; and

    (b)   in the case of an outside worker employed by an employer in Northern Ireland or in another member State, a passbook authorised by the competent authority for Northern Ireland or that member State, as the case may be;

"radiation protection adviser" means, subject to regulation 39(6) (which relates to transitional provisions), an individual who, or a body which, meets such criteria of competence as may from time to time be specified in writing by the Executive;

4   The Health and Safety Executive has published a statement[8] specifying the criteria of competence for basic Radiation Protection Adviser capability. This statement is available from HSE Radiation Protection Policy, Health Directorate, Rose Court, 2 Southwark Bridge, London SE1 9HS. It is also reproduced on the Health Directorate page of HSE's website. The criteria cover both individuals and bodies that wish to give advice as radiation protection advisers (RPAs). An individual wishing to be accepted as an RPA should either hold a valid certificate from an assessing body recognised by HSE or have received a national vocational qualification or Scottish vocational qualification (N/SVQ) at level 4 in Radiation Protection Practice within the previous five years. RPA bodies need to apply to HSE for formal recognition. The HSE statement includes specific requirements that assessing bodies have to meet to be accepted by HSE for the purpose of assessing the competence of individuals to act as RPAs.

5   Certificates awarded by assessing bodies must state that the individual has satisfied the assessing body that he or she possesses the core competence needed to be an RPA. A certificate will be valid for no more than five years. The certificate must be of general application, ie it cannot be limited to work in one sector. However, assessing bodies may choose to acknowledge specific areas of expertise that the individual possesses. These acknowledgements will often reflect the fact that the individual has gained much of their knowledge and experience in specific work sectors. Such acknowledgements should be of assistance to radiation employers when selecting 'suitable' RPAs (see guidance on regulation 13). However, they are not required as part of the HSE criteria of core competence.

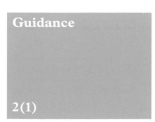

Guidance

6   Where an individual satisfies the HSE criteria by possession of an N/SVQ, they will need to apply to an assessing body for a certificate within five years from the date of issue. This process is the equivalent to renewing a certificate from an assessing body by showing that professional knowledge and experience have been maintained by the individual since receiving the N/SVQ.

7   For organisations wishing to act as RPAs (RPA bodies), the essential elements of the criteria of competence are that the personnel of the RPA body should include at least one individual RPA (and normally more) with a valid certificate from an assessing body (or N/SVQ) and that all advice emanating from the RPA body is traceable, through management systems and quality assurance procedures, to such certificated RPA(s). Organisations wishing to act as RPA bodies have to apply to HSE for appropriate recognition.

8   To allow for the change-over to this competence-led system, regulation 39(6) contains a five-year transitional provision. During the period up to the end of 2004, any individual or organisation that can show they have held an appointment as an RPA under the Ionising Radiations Regulations 1985 can act as an RPA under the new Regulations. New RPAs, though, will need to meet the criteria of competence from the outset.

2(1)

Regulation

2(1)

"radioactive substance" means any substance which contains one or more radionuclides whose activity cannot be disregarded for the purposes of radiation protection;

ACOP

2(1)

9   **For a substance used in a practice, its activity should never be disregarded for the purposes of radiation protection where that activity exceeds the values set out in column 2 of Schedule 8, subject to the quantity of the substance also exceeding the values set out in column 3 of Schedule 8.**

Guidance

10   Whether a particular material constitutes a radioactive substance within the meaning of the Regulations depends on a judgement about its radiological impact. For substances used in practices, the Approved Code of Practice gives practical advice on situations where substances should always be regarded as radioactive substances. Below these activity levels, substances may still be considered 'radioactive' for the purposes of the Regulations, depending on the radionuclide concerned, the way in which they are used and the particular circumstances of exposure. For example, a substance used only in solid form may have an activity that can be disregarded, but the same substance used in such a way that people can inhale that material as a fine dust might need to be treated as a radioactive substance for the purposes of these Regulations.

2(1)

ACOP

2(1)

11   **In the special case of substances containing naturally occurring radionuclides used in work other than a practice, their activity cannot be disregarded for the purposes of radiation protection where their use is likely to lead to employees or other people receiving an effective dose of ionising radiations in excess of 1 millisievert in a year.**

Guidance

2(1)

12   The judgement about what constitutes a radioactive substance within the meaning of the Regulations can be particularly difficult for materials containing naturally occurring radionuclides. Processing of such materials can lead to concentration of certain of these radionuclides and to the potential for significant exposure at particular stages of a process. There are two main exposure routes: direct exposure to external radiation from bulk quantities, often held in store; and inhalation arising from dusty operations.

13   An initial assessment should indicate whether it is likely that exposure could lead to anyone receiving an effective dose approaching or exceeding 1 mSv a year. If this appears to be the case, a more detailed assessment might be necessary to confirm that the material should be considered radioactive for regulatory purposes, to decide where the risk arises and to determine what protective measures might be needed (see regulation 7). Some processes with a recognised potential to cause significant exposure are:

(a)   oil and gas extraction, where scale in pipes and vessels may contain significant amounts of uranium and thorium and their decay products including radium;

(b)   some forms of metal, where refractory materials or feed ores may contain naturally occurring radioactive materials and where radionuclides can volatilise and condense, or concentrate in the product or the slags to enhance activity levels; and

(c)   thorium alloy manufacture, for example for aircraft parts, and the use of thoriated products, such as special types of welding electrodes.

"sealed source" means a source containing any radioactive substance whose structure is such as to prevent, under normal conditions of use, any dispersion of radioactive substances into the environment, but it does not include any radioactive substance inside a nuclear reactor or any nuclear fuel element;

"short-lived daughters of radon 222" means polonium 218, lead 214, bismuth 214 and polonium 214;

"supervised area" means an area which has been so designated by the employer in accordance with regulation 16(3);

"trainee" means a person aged 16 years or over (including a student) who is undergoing instruction or training which involves operations which would, in the case of an employee, be work with ionising radiation;

"transport" means, in relation to a radioactive substance, carriage of that substance on a road within the meaning of, in relation to England and Wales, section 192 of the Road Traffic Act 1988[(a)] and, in relation to Scotland, the Roads (Scotland) Act 1984[(b)] or through another public place (whether on a conveyance or not), or by rail, inland waterway, sea or air and, in the case of transport on a conveyance, a substance shall be deemed as being transported from the time that it is loaded onto the conveyance for the purpose of transporting it until it is unloaded from that conveyance, but a substance shall not be considered as being transported if -

(a)   it is transported by means of a pipeline or similar means; or

(b)   it forms an integral part of a conveyance and is used in connection with the operation of that conveyance;

"woman of reproductive capacity" means a woman who is made subject to the additional dose limit for a woman of reproductive capacity specified in paragraphs 5 and 11 of Schedule 4 by an entry in her health record made by an appointed doctor or employment medical adviser;

(a) 1988 c.52.
(b) 1984 c.54.

"work with ionising radiation" means work to which these Regulations apply by virtue of regulation 3(1).

14   Essentially, work with ionising radiation means:

(a)   a practice, which involves the production, processing, handling, use, holding, storage, transport or disposal of artificial radioactive substances and some naturally occurring sources, or the use of electrical equipment emitting ionising radiation at more than 5 kV (see definition of practice in regulation 2(1));

(b)   work in places where the radon gas concentration exceeds the values in regulation 3(1)(b); or

(c)   work with radioactive substances containing naturally occurring radionuclides not covered by the definition of a practice (see paragraph 11).

(2)   *In these Regulations, unless the context otherwise requires, any reference to -*

(a)   *an employer includes a reference to a self-employed person and any duty imposed by these Regulations on an employer in respect of his employee shall extend to a self-employed person in respect of himself;*

(b)   *an employee includes a reference to -*

(i)   *a self-employed person, and*

(ii)   *a trainee who but for the operation of this sub-paragraph and paragraph (3) would not be classed as an employee;*

(c)   *exposure to ionising radiation is a reference to exposure to ionising radiation arising from work with ionising radiation;*

(d)   *a person entering, remaining in or working in a controlled or supervised area includes a reference to any part of a person entering, remaining in or working in any such area.*

15   A self-employed person who works with ionising radiation will simultaneously have certain duties under these Regulations both as an employer and as an employee (see guidance to regulation 34 and Introduction).

(3)   *For the purposes of these Regulations and Part I of the Health and Safety at Work etc. Act 1974 -*

(a)   *the word "work" shall be extended to include any instruction or training which a person undergoes as a trainee and the meaning of "at work" shall be extended accordingly; and*

(b)   *a trainee shall, while he is undergoing instruction or training in respect of work with ionising radiation, be treated as the employee of the person whose undertaking (whether for profit or not) is providing that instruction or training and that person shall be treated as the employer of that trainee except that the duties to the trainee imposed upon the person providing instruction or training shall only extend to matters under the control of that person.*

(4) In these Regulations, where reference is made to a quantity specified in Schedule 8, that quantity shall be treated as being exceeded if –

(a) where only one radionuclide is involved, the quantity of that radionuclide exceeds the quantity specified in the appropriate entry in Schedule 8; or

(b) where more than one radionuclide is involved, the quantity ratio calculated in accordance with Part II of Schedule 8 exceeds one.

(5) Nothing in these Regulations shall be construed as preventing a person from entering or remaining in a controlled area or a supervised area where that person enters or remains in any such area –

(a) in the due exercise of a power of entry conferred on him by or under any enactment; or

(b) for the purpose of undergoing a medical exposure.

(6). In these Regulations –

(a) any reference to an effective dose means the sum of the effective dose to the whole body from external radiation and the committed effective dose from internal radiation; and

(b) any reference to equivalent dose to a human tissue or organ includes the committed equivalent dose to that tissue or organ from internal radiation.

(7) In these Regulations –

(a) a numbered Regulation or Schedule is a reference to the Regulation or Schedule in these Regulations so numbered;

(b) a numbered paragraph is a reference to the paragraph so numbered in the Regulation or Schedule in which that reference appears.

Regulation 3

Application

(1) Subject to the provisions of this regulation and to regulation 6(1), these Regulations shall apply to –

(a) any practice;

(b) any work (other than a practice) carried out in an atmosphere containing radon 222 gas at a concentration in air, averaged over any 24 hour period, exceeding 400 Bq m^{-3} except where the concentration of the short-lived daughters of radon 222 in air averaged over any 8 hour working period does not exceed 6.24×10^{-7} Jm^{-3}; and

(c) any work (other than work referred to in sub-paragraphs (a) and (b) above) with any radioactive substance containing naturally occurring radionuclides.

16   This paragraph sets out the three basic types of work to which the Regulations apply. A practice, the most common type, is defined in regulation 2(1) and covers the normal activities associated with use of radiation sources, such as medical uses, power generation and industrial radiography. A practice also includes some forms of work with materials containing naturally occurring

radionuclides. Regulation 3(1)(b) covers work in mines and other workplaces in certain radon-prone areas of Great Britain where the construction of the workplace and the ventilation provided is insufficient to keep the concentration below the specified levels. Regulation 3(1)(c) will only apply where the work involves a radioactive substance as explained in paragraph 11. In most instances, controls are the same for all three types but there is some differentiation, for example in regulation 3(2) and regulation 6(7).

(2)   The following Regulations shall not apply where the only work being undertaken is that referred to in sub-paragraph (b) of paragraph (1), namely regulations 23, 27 to 30, 32 and 33.

(3)   The following regulations shall not apply in relation to persons undergoing medical exposures, namely regulations 7, 8, 11, 16 to 18, 23, 25, 31(1) and 34(1).

(4)   Regulation 11 shall not apply in relation to any comforter and carer.

(5)   In the case of an outside worker (working in a controlled area situated in Great Britain) employed by an employer established in Northern Ireland or in another member State, it shall be sufficient compliance with regulation 21 (dose assessment and recording) and regulation 24 (medical surveillance) if the employer complies with -

(a)   where the employer is established in Northern Ireland, regulations 13 and 16 of the Ionising Radiations Regulations (Northern Ireland) 1985[a]; or

(b)   where the employer is established in another member State, the legislation in that State implementing Chapters II and III of Title VI of the Directive where such legislation exists.

(a) SR 1985/273.

17   This requirement avoids the need for a further pre-assignment medical examination when an outside worker arrives in Great Britain from another member state (or Northern Ireland). It also means that an outside worker from another member state does not have to wear an additional dosemeter, issued by an Approved Dosimetry Service (ADS) in Great Britain, for the purpose of dose assessment.

Duties under the Regulations

(1)   Any duty imposed by these Regulations on an employer in respect of the exposure to ionising radiation of persons other than his employees shall be imposed only in so far as the exposure of those persons to ionising radiation arises from work with ionising radiation undertaken by that employer.

(2)   Duties under these Regulations imposed upon the employer shall also be imposed upon -

(a)   the manager of a mine (within the meaning of section 180 of the Mines and Quarries Act 1954[a]); and

(b)   the operator of a quarry (within the meaning of the Quarries Regulations 1999[b]),

in so far as those duties relate to the mine or part of the mine of which he is the manager or the quarry of which he is the operator and to matters within his control.

(3)   Subject to regulation 6(1)(b), duties under these Regulations imposed upon the employer shall also be imposed on the holder of a nuclear site licence under the Nuclear Installations Act 1965[c] in so far as those duties relate to the licensed site.

(a) 1954 c.70; section 180 was amended by SI 1993/1897.
(b) SI 1999/2024.
(c) 1965 c.57; relevant amending instruments are SI 1974/2056 and SI 1990/1918.

PART II GENERAL PRINCIPLES AND PROCEDURES

Authorisation of specified practices

(1)   Subject to paragraph (2), a radiation employer shall not, except in accordance with a prior authorisation granted by the Executive in writing for the purposes of this paragraph, carry out the following practices -

(a)   the use of electrical equipment intended to produce x-rays for the purpose of -

(i)   industrial radiography;

(ii)   the processing of products;

(iii)   research; or

(iv)   the exposure of persons for medical treatment; or

(b)   the use of accelerators, except electron microscopes.

(2)   Paragraph (1) shall not apply in respect of any practice of a type which is for the time being authorised by the Executive where such practice is or is to be carried out in accordance with such conditions as may from time to time be approved by the Executive in respect of that type of practice.

(3)   An authorisation granted under paragraph (1) may be granted subject to conditions and with or without limit of time and may be revoked in writing at any time.

18   Employers wishing to undertake one of the practices listed in regulation 5(1) must either do so in accordance with the specified conditions in a generic authorisation (regulation 5(2)) or must obtain individual prior authorisation from HSE. This requirement applies both to employers undertaking a specified practice for the first time and to those who started such work before this regulation came into force on 13 May 2000. Only the practices specified in the regulation need prior authorisation under these Regulations.

19   The use of electrical equipment to produce X-rays (X-ray sets) for industrial radiography is interpreted as involving non-destructive testing of welds or structural integrity and therefore excludes forms of routine surveillance or analytical systems such as the use of baggage, postal or food screening. Processing of products is regarded as inducing some physical, chemical or biological change in the item or material being processed. It does not, therefore, include non-destructive testing, measurement or gauging systems, or X-ray analytical techniques. The use of X-ray sets for research includes some industrial, academic and medical uses. An example would be the use of an X-ray set to deliver high doses to laboratory animals in

radiobiology research. In this context, research covers creative work undertaken on a systematic basis in order to increase the stock of knowledge. It includes basic research, undertaken primarily to acquire new knowledge of the underlying foundation of phenomena and observable facts, and applied research (ie original investigation directed primarily towards a specific practical aim or objective). The use of X-ray sets for routine analytical, medical and veterinary diagnostic or investigational purposes, except as part of such a research project, is not regarded as being subject to prior authorisation.

20   An accelerator is defined in regulation 2(1). Uses include non-destructive testing, industrial, academic and medical research and development, medical treatment and radionuclide manufacture.

Generic authorisations

21   HSE has issued generic authorisations covering each of the practices in regulation 5(1) (copies may be obtained from HSE local offices or downloaded from HSE's website). The conditions incorporated in the generic authorisations are designed to ensure that any risk of exposure to ionising radiation arising from these practices is limited. Radiation employers who rely on generic authorisation must comply with those conditions, but this is only one step that must be taken before beginning a practice listed in regulation 5(1). The radiation employer must also comply with all other relevant duties under these Regulations.

Individual prior authorisation

22   If employers cannot comply fully with the conditions in the relevant generic authorisation, they should apply to HSE for individual authorisation before undertaking a practice listed in regulation 5(1). An information sheet[9] is available from HSE offices on how to apply for an individual authorisation. HSE will charge a fee for granting individual prior authorisations.

Registration and authorisation under the Radioactive Substances Act 1993

23   Radiation employers who use both radioactive substances and X-ray sets may also have to seek registration and authorisation under the Radioactive Substances Act 1993[6] (RSA93), from the Environment Agencies, for the keeping and disposal of radioactive substances. An example would be use of an X-ray pipeline crawler that has a radionuclide control source. The pipeline crawler X-ray set would require prior authorisation under these Regulations and the source would require authorisation under RSA93. HSE and the Agencies will co-ordinate their activities as far as possible to avoid or reduce inconsistency.

Notifying changes to HSE

(4)   Where an authorisation has been granted pursuant to paragraph (1) and the radiation employer to whom the authorisation was granted subsequently makes a material change to the circumstances relating to that authorisation, that change shall forthwith be notified to the Executive by the radiation employer.

24   Radiation employers who have had to obtain individual authorisations under these Regulations will need to tell HSE if the work they do changes to such an extent that the particulars relating to the authorisation become incorrect or out of date.

Appeals

(5)   A radiation employer to whom this regulation applies and who is aggrieved by -

(a)   a decision of the Executive -

(i)   refusing to grant an authorisation under paragraph (1);

(ii)   imposing a limit of time upon an authorisation granted under paragraph (1); or

(iii)   revoking an authorisation under paragraph (3); or

(b)   the terms of any condition attached to the authorisation by the Executive under paragraph (3),

may appeal to the Secretary of State.

25   HSE would advise radiation employers how to make such an appeal, if the occasion arose.

(6)   Sub-sections (2) to (6) of section 44 of the 1974 Act shall apply for the purposes of paragraph (5) as they apply to an appeal under section 44(1) of that Act.

(7)   The Health and Safety Licensing Appeals (Hearings Procedure) Rules 1974[(a)] as respects England and Wales, and the Health and Safety Licensing Appeals (Hearings Procedure) (Scotland) Rules 1974[(b)], as respects Scotland, shall apply to an appeal under paragraph (5) as they apply to an appeal under sub-section (1) of the said section 44, but with the modification that references to a licensing authority are to be read as references to the Executive.

(a) SI 1974/2040.
(b) SI 1974/2068.

26   Regulation 5(5) provides for an appeal by a radiation employer against a decision by HSE about an authorisation or conditions attached to an authorisation. Regulation 5(6) provides that such an appeal is treated as if it were an appeal under section 44 of the Health and Safety at Work etc Act 1974[2] for the purpose of conferring powers on the Secretary of State in relation to the determination of the appeal. Regulation 5(7) provides that the rules governing appeals hearings in England and Wales and Scotland also apply to such an appeal.

Regulation 6

Notification of specified work

(1)   This regulation shall apply to work with ionising radiation except -

(a)   work specified in Schedule 1; and

(b)   work carried on at a site licensed under section 1 of the Nuclear Installations Act 1965.

Exemptions from notification

27   Schedule 1 specifies work with ionising radiation that does not need to be notified to HSE. This Schedule refers, in particular, to quantities and concentrations set out in Schedule 8, at or below which notification is not required. Schedule 8 covers all radionuclides likely to be used in Great Britain.

HSE has power to approve specific values for unlisted radionuclides. Where this has not been done, the Schedule provides default values. Part II of Schedule 8 explains how to calculate values for exemption from notification where more than one radionuclide is involved.

28   Work with ionising radiation on licensed nuclear sites is not subject to notification under this regulation (although it may be subject to prior authorisation under regulation 5).

Notification to HSE and exemptions

(2)   Subject to paragraphs (7) and (8) and to regulation 39(1) (which relates to transitional provisions), a radiation employer shall not for the first time carry out work with ionising radiation to which this regulation applies unless at least 28 days before commencing that work or before such shorter time as the Executive may agree he has notified the Executive of his intention to carry out that work and has provided the Executive with the particulars specified in Schedule 2.

(3)   Where a radiation employer has notified work in accordance with paragraph (2), the Executive may, by notice in writing served on him, require that radiation employer to provide such additional particulars of that work as it may reasonably require, being any or all of the particulars specified in Schedule 3, and in such a case the radiation employer shall provide those particulars by such time as is specified in the notice or by such other time as the Executive may subsequently agree.

29   Unless exempted by regulation 6(1), radiation employers will generally need to send HSE a notification of intention to work 28 days before starting work with ionising radiation (but see paragraph 35 for work notified to HSE under previous Regulations). Schedule 2 sets out the information that the employer should send to HSE. From time to time, HSE will say in what formats it is willing to accept such information (post, fax and e-mail are currently acceptable). In general, radiation employers should notify the HSE office local to the place where it is intended to undertake the work with ionising radiation.

30   Regulation 39(1) is a transitional provision covering work with ionising radiation first undertaken between 1 January and 26 February 2000.

31   Employers who are subject to prior authorisation under regulation 5 will also need to notify HSE under regulation 6 (but see paragraph 28).

32   Where the work with ionising radiation involves a substance containing naturally occurring radionuclides and falls within the definition of a practice (see regulation 2(1)), then the requirements for notification apply unless Schedule 1 provides exemption. If the work is not a practice as defined, but the Regulations apply by virtue of regulation 3(1)(b) or (c), notification may be retrospective (see regulation 6(7)).

Special notification to HSE for each job

(4)   A notice under paragraph (3) may require the radiation employer to notify the Executive of any of the particulars specified in Schedule 3 before each occasion on which he commences work with ionising radiation.

Notifying changes to HSE

(5)   Where a radiation employer has notified work in accordance with paragraph (2) and subsequently makes a material change in that work which would affect the particulars so notified, he shall forthwith notify the Executive of that change.

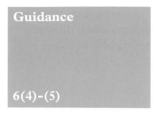

Guidance

6(4)-(5)

33   'Material change' will have the same meaning as in regulation 5(4). Radiation employers should tell HSE if the particulars previously notified are no longer correct, such as: if the employer's particulars or particulars of the premises change; if the category of source changes; if it is decided to use a source at a different premises. For example, if the original notification covered the use of an X-ray set for veterinary purposes and it was decided to start using radioactive substances as well, this change would need to be notified.

Regulation

6(6)

(6)   Nothing in paragraph (5) shall be taken as requiring the cessation of the work to be notified in accordance with that paragraph except where the site or any part of the site in which the work was carried on has been or is to be vacated.

Regulation

6(7)

Notification of work in radon atmospheres or with substances containing naturally occurring radionuclides

(7)   Where the only work being undertaken is work referred to in regulation 3(1)(b) or (c), it shall be a sufficient compliance with paragraph (2) if the radiation employer having control of the premises where the work is carried on makes the notification required by that paragraph forthwith after the work has commenced.

Guidance 6(7)

34   See guidance on regulation 6(2).

Regulation

Notification by nursing homes and other places for the first time after patients have been administered with a radioactive medicinal product

(8)   In relation to work involving the care of a person to whom a radioactive medicinal product (within the meaning of the Medicines (Administration of Radioactive Substances) Regulations 1978[a] has been administered, it shall be sufficient compliance with paragraph (2) if the notification required by that paragraph is given as soon as is practicable before the carrying out of that work.

(a) SI 1978/1006.

Notifications made before the coming into force of these Regulations

(9)   Where in respect of work with ionising radiation carried out prior to the coming into force of these Regulations notification has been given to the Executive pursuant to any statutory requirement, the provisions of this regulation shall apply to such notification as if that notification had been given in accordance with paragraph (2).

6(8)-(9)

Guidance

6(8)-(9)

35   This means that it is not necessary to repeat a notification properly made (or deemed to have been made) to HSE under the Ionising Radiations Regulations 1985 (IRR85).

Regulation 7

Prior risk assessment etc

Regulation

(1)   Before a radiation employer commences a new activity involving work with ionising radiation in respect of which no risk assessment has been made by him, he shall make a suitable and sufficient assessment of the risk to any employee and other person for the purpose of identifying the measures he needs to take to restrict the exposure of that employee or other person to ionising radiation.

(2)   Without prejudice to paragraph (1), a radiation employer shall not carry out work with ionising radiation unless he has made an assessment sufficient to demonstrate that -

(a)   all hazards with the potential to cause a radiation accident have been identified; and

(b)   the nature and magnitude of the risks to employees and other persons arising from those hazards have been evaluated.

7

20

(3)   Where the assessment made for the purposes of this regulation shows that a radiation risk to employees or other persons exists from an identifiable radiation accident, the radiation employer shall take all reasonably practicable steps to -

(a)   prevent any such accident;

(b)   limit the consequences of any such accident which does occur; and

(c)   provide employees with the information, instruction and training, and with the equipment necessary, to restrict their exposure to ionising radiation.

(4)   The requirements of this regulation are without prejudice to the requirements of regulation 3 (Risk assessment) of the Management of Health and Safety at Work Regulations 1992[(a)].

(a) SI 1992/2051.

General advice on risk assessment

36   This requirement for a prior risk assessment complements the related requirements of regulation 3 of the Management of Health and Safety at Work Regulations 1999[7] (MHSWR) which updated and superseded the 1992 Management Regulations. HSE publications giving guidance specifically on MHSWR are available from HSE Books, see References at the end of this publication.

37   The essential difference in regulation 7(1) is that a new activity involving work with ionising radiation may not begin until the risk assessment has been made. Once the work commences, regulation 3 of MHSWR requires the recording of the assessment (if there are five or more employees) and the maintenance of the risk assessment to keep it up to date where there has been a significant change in the matters to which it relates. Regulation 5 of MHSWR also requires the making of arrangements for effective planning, organisation, control, monitoring and review of preventive and protective measures.

38   If the intended activity is already covered by a suitable and sufficient assessment which was undertaken for the purposes of MHSWR, nothing further needs to be done to satisfy regulation 7(1) of these Regulations. The radiation employer is not expected to carry out a further risk assessment on each occasion before the activity starts if the working conditions are unchanged (see paragraph 46).

39   Where the work with ionising radiation was being carried out before these Regulations come into force and the employer has carried out a risk assessment as required by regulation 3 of MHSWR, that assessment may need to be reviewed to make sure it remains sufficient and suitable. The purpose of the risk assessment under MHSWR is to help the employer or self-employed person determine what measures should be taken to comply with relevant duties under the 'relevant statutory provisions', which include health and safety regulations such as these Regulations. New legal duties on employers (including radiation employers) have been introduced by these Regulations. Therefore, employers might find that the original assessment has to be revised. One example would be the need to consider any necessary changes to the working conditions of a pregnant female employee - regulation 8(5) is more specific than regulations 16 to 18 of MHSWR in relation to risks to new or expectant mothers.

40   A suitable and sufficient prior risk assessment made under regulation 7(1) for any new activities will be sufficient to satisfy the requirements of regulation 3 of MHSWR as far as radiation protection is concerned. However, the radiation employer will not want to consider the radiation protection aspects of the work in isolation from other health and safety considerations, for example: some control methods for restricting exposure to ionising radiation by use of distance and shielding might pose unacceptable risks of falls or back strain. Therefore, the radiation employer will need to consider the differing radiological and conventional risks associated with alternative techniques under consideration for the work, in order to satisfy both regulation 7 of these Regulations and regulation 3 of MHSWR.

41   Employers who undertake work other than work with ionising radiation (eg general maintenance or cleaning) at the premises of a radiation employer may need to take account of matters identified in the radiation employer's risk assessment when making or reviewing their own risk assessment to satisfy MHSWR.

42   Regulation 7(1) does not apply to the protection of those undergoing a medical examination or treatment. However, it does apply to the protection of staff who carry out those exposures to members of the public and to 'comforters and carers' as defined in regulation 2(1).

Responsibility for undertaking the prior risk assessment

43   The duty to undertake a prior risk assessment is placed on the radiation employer since, for the purposes of this regulation, a radiation employer is defined as including an employer who intends to carry out work with ionising radiation (see paragraph 3).

Nature of prior risk assessment for new activities

44   Where a radiation employer is required to undertake a prior risk assessment, the following matters need to be considered, where they are relevant:

(a)   the nature of the sources of ionising radiation to be used, or likely to be present, including accumulation of radon in the working environment;

(b)   estimated radiation dose rates to which anyone can be exposed;

(c)   the likelihood of contamination arising and being spread;

(d)   the results of any previous personal dosimetry or area monitoring relevant to the proposed work;

(e)   advice from the manufacturer or supplier of equipment about its safe use and maintenance;

(f)   engineering control measures and design features already in place or planned;

(g)   any planned systems of work;

(h)   estimated levels of airborne and surface contamination likely to be encountered;

(i)   the effectiveness and the suitability of personal protective equipment to be provided;

(j)   the extent of unrestricted access to working areas where dose rates or contamination levels are likely to be significant;

(k)   possible accident situations, their likelihood and potential severity;

(l)   the consequences of possible failures of control measures - such as electrical interlocks, ventilation systems and warning devices - or systems of work;

(m)   steps to prevent identified accidents situations, or limit their consequences.

45   This prior risk assessment should enable the employer to determine:

(a)   what action is needed to ensure that the radiation exposure of all persons is kept as low as reasonably practicable (regulation 8(1));

(b)   what steps are necessary to achieve this control of exposure by the use of engineering controls, design features, safety devices and warning devices (regulation 8(2)(a)) and, in addition, by the development of systems of work (regulation 8(2)(b));

(c)   whether it is appropriate to provide personal protective equipment and if so what type would be adequate and suitable (regulation 8(2)(c));

(d)   whether it is appropriate to establish any dose constraints for planning or design purposes, and if so what values should be used (regulation 8(3));

(e)   the need to alter the working conditions of any female employee who declares she is pregnant or is breastfeeding (regulation 8(5));

(f)   an appropriate investigation level to check that exposures are being restricted as far as reasonably practicable (regulation 8(7));

(g)   what maintenance and testing schedules are required for the control measures selected (regulation 10);

(h)   what contingency plans are necessary to address reasonably foreseeable accidents (regulation 12);

(i)   the training needs of classified and non-classified employees (regulation 14);

(j)   the need to designate specific areas as controlled or supervised areas and to specify local rules (regulations 16 and 17);

(k)   the actions needed to ensure restriction of access and other specific measures in controlled or supervised areas (regulation 18);

(l)   the need to designate certain employees as classified persons (regulation 20);

(m) the content of a suitable programme of dose assessment for employees designated as classified persons and for others who enter controlled areas (regulations 18 and 21);

(n) the responsibilities of managers for ensuring compliance with these Regulations; and

(o) an appropriate programme of monitoring or auditing of arrangements to check that the requirements of these Regulations are being met.

46   Where the employer conducts a number of very similar activities, such as radiography operations, then a generic assessment would generally be sufficient provided it encompassed the range of risks that are likely to be encountered.

47   Paragraph 217 advises the radiation employer to consult a radiation protection adviser (RPA) about the matters to be considered when conducting a prior risk assessment. Also, it may be necessary to consult the appointed safety representative(s), and, where appropriate, any established safety committee, about the introduction of new measures at the workplace which may affect health and safety (regulation 4(a) of the Safety Representatives and Safety Committees Regulations 1977).[10] Where there is no appointed safety representative the employer may need to consult employees, as required by the Health and Safety (Consultation with Employees) Regulations 1996.[11]

48   As part of the prior risk assessment, radiation employers will need to take account of the risks arising from radiation exposure of expectant mothers and mothers who are breastfeeding, and in particular the likely doses to the foetus or breastfed infant. Also, the assessment should, where appropriate, take into account the particular risks to young people resulting from their inexperience, lack of awareness of risks and possible immaturity (regulation 3(4) MHSWR).[7]

49   The prior risk assessment might provide a suitable basis for helping to establish dose constraints, where these are appropriate (see regulation 8(3)). The radiation employer may take account of information about past operating experience and recommendations from relevant professional bodies or trade associations in establishing dose constraints.

Prior risk assessment for simple uses of ionising radiation

50   The prior risk assessment is only a tool to help the radiation employer decide on the most appropriate control measures for the new work activities. Completing the prior risk assessment should be a straightforward process for employers involved in simple uses of ionising radiation, for example the use of a nucleonic level or thickness gauge. In these cases, the radiation protection issues are not likely to be complex. Frequently, the assessment will require nothing more than common sense judgement based on the application of advice provided by the manufacturer or supplier of the device intended for use. In particular, the assessment is likely to be straightforward where:

(a) the employer follows the clearly stated advice of the RPA; or

(b) the employer decides to adopt controls and working procedures which adhere to accepted standards in the industry or are contained in authoritative codes of practice or guidance; or

(c) the work involves the use of a piece of equipment which has been well designed to keep exposures as low as reasonably practicable, and which is functioning properly, and the employer needs only to develop procedures for the safe use of the equipment (see regulation 7(2) for considering possible incidents arising from a malfunction of the equipment). Advice on the safe use and maintenance of equipment should be available from the manufacturer or supplier.

Recording the results of the prior risk assessment

51 Employers with five or more employees have a duty under regulation 3(6) of MHSWR[7] to record the significant findings of their risk assessment, and any group of employees identified as being especially at risk. It therefore makes sense for the prior risk assessment to be similarly recorded. This record, which may be electronic, will represent an effective statement of the risks the work presents and will lead management to take the necessary actions to protect employees and others exposed to ionising radiation. Where appropriate, assessment records could be linked with other health and safety records such as:

(a) the health and safety arrangements required by regulation 5 of MHSWR;

(b) the written health and safety policy statement required by section 2(3) of the Health and Safety at Work etc Act 1974[2] (HSW Act); and

(c) any local rules required by regulation 17(1) of these Regulations.

Review and revision of the prior risk assessment

52 Regulation 3(3) of MHSWR requires the employer to review the assessment if there is reason to suspect it is no longer valid or there has been a significant change in the work activity. However, in most cases it will be prudent to review the validity of the risk assessment and the correctness of its conclusions periodically, as part of standard health and safety management practice. In general, employers should normally decide the frequency of reviews by taking account of the nature of the work, the degree of risk and the extent of any likely change in the work activity.

53 One way in which the employer might discover that an assessment is invalid is through checking the results of personal dosimetry or area monitoring. These results could indicate a breakdown of controls and so highlight the need for a formal review of whether the procedures in place are satisfactory.

54 A significant change in the work activity may include such matters as:

(a) the introduction of a radioactive source of a much larger activity or a source which emits a different type or quality of radiation;

(b) the use of electrical equipment which produces X-rays of much higher energy;

(c) the introduction of unsealed sources in an area where only sealed sources have previously been used;

(d) plant modification, including alterations to engineering controls and safety features;

(e) changes to the process or methods of work;

(f) human factors, eg arising from staff turnover.

55   Further advice on the review of risk assessments is given in HSE publications on MHSWR[7] available from HSE Books (see References).

Assessments for accident hazards

56   Regulation 7(2) specifically requires employers to assess the work with ionising radiation they intend to undertake for possible radiation accidents (defined in regulation 2(1)), which will normally mean establishing what accident scenarios are possible, their likelihood and their potential severity. Where the work was in progress before the coming into force of these Regulations, the radiation employer will need to ensure that any assessment of hazards carried out in compliance with IRR85 remains sufficient for the purposes of regulation 7(2) of these Regulations.

57   These assessments need to take account of the consequences not only of possible plant and equipment failures but also of a breakdown in work systems and predictable forms of unauthorised behaviour at work. The scope and comprehensiveness of this aspect of the assessment should match the circumstances. If a particular accident scenario is shown to be either extremely unlikely or trivial in its consequences, then the assessment need go no further.

58   Once the assessment has identified how an accident could occur, regulation 7(3) requires reasonably practicable measures either to prevent its happening or to limit its consequences. These measures need to be permanent in nature to achieve an ongoing reduction of risk and are different from the planned actions designed to mitigate an accident once it takes place which are likely to be reflected in the contingency plan required by regulation 12. They flow naturally from the analysis of accident causation in the assessment. Certain employers holding more than specified amounts of radioactive substances will also need to consider the requirements of regulation 26 IRR85, (special hazard assessments) until it is replaced by separate regulations on emergency preparedness (see regulation 41(3)).

Restriction of exposure

(1)   Every radiation employer shall, in relation to any work with ionising radiation that he undertakes, take all necessary steps to restrict so far as is reasonably practicable the extent to which his employees and other persons are exposed to ionising radiation.

59   Dose sharing should not be used as a primary means of keeping exposures below the dose limits. Rather, the radiation employer should give priority to improving engineering controls and adopting other means of restricting exposure, including changing the methods of work. However, if a choice has to be made between restricting doses to individuals and restricting doses to a group of persons, priority should be given to keeping individual doses as far below dose limits as is reasonably practicable.

60   Radiation employers should take particular steps to restrict the exposure of any employees who would not normally be exposed to ionising radiation in the course of their work. The dose control measures should make it unlikely that such persons would receive an effective dose greater than 1 millisievert per year or an equivalent dose which exceeds that specified as a dose limit for any other person in Schedule 4.

Responsibility for restricting exposure

61   The radiation employer, as defined by regulation 2(1), has the overall responsibility for restricting exposure to ionising radiation and will need to co-operate with any other employers whose employees are affected by the work with ionising radiation, as required by regulation 15 (see paragraphs 242-247 about co-operation generally and 312-313 about outside workers).

General approach to restricting exposure

62   This general duty to restrict exposure covers both the decision to work with ionising radiation and the selection of any source of ionising radiation for the work. One of the first considerations may be whether the risk can be avoided by choosing an alternative technique which does not involve ionising radiation. However, any alternative techniques considered which avoid work with ionising radiation should not lead to an overall increase in the health and safety risks to which employees and other people will be exposed.

63   In general, the lower the activity of any radioactive source used, commensurate with the work that needs to be done, the easier it will be to ensure that exposure of employees and others is adequately restricted. Similar consideration can be given to the minimisation of unnecessary radiation when X-ray sets are used. This may involve the use of appropriate beam filtration, effective collimation of the X-ray beam and a careful choice of the operating voltage and tube current.

64   Where the design of new facilities is being considered for work with ionising radiation, the radiation employer will need to consider the construction, commissioning and operation of the facility together with its maintenance and decommissioning to ensure that exposure will be restricted as far as reasonably practicable during its life.

65   Radiation employers can obtain valuable feedback about the success of their control measures to restrict exposure by analysing monitoring data collected in accordance with regulations 18(7)(d), 19 and 21. Regulation 19(1) specifically requires that the working conditions in controlled and supervised areas are kept under review. Radiation employers will need to think about how best to review the available information (and at what frequency) to identify any further action that may be required under this regulation. It may be appropriate to supplement the information with task-specific dose measurements in some situations, for example where doses to individuals might be a significant fraction of a relevant dose limit. A suitable RPA appointed under regulation 13 would normally be able to advise the radiation employer about collecting and analysing relevant information. Appointed safety representatives (and, where appropriate, an established safety committee or dose reduction group) might also need to be consulted about reviews of the data.

66   If there are significant variations in doses for the same individual(s) between different periods in the year, or between different individuals undertaking similar work, it would be prudent to undertake a more detailed review. Where the dose for an individual exceeds the investigation level established under regulation 8(7), the employer of the person concerned has to undertake a formal investigation (see paragraphs 152-162).

Examination of plans and acceptance testing

67   Where new equipment or apparatus is provided or installed, the radiation employer would normally arrange for acceptance testing of that equipment or

27

apparatus to make sure that it conforms to specification. In accordance with regulation 13 and Schedule 5, the RPA must be consulted about the prior examination of plans and acceptance testing of new or modified sources of ionising radiation and the adequacy (and functioning) of control measures provided to restrict exposure. Also, it may be necessary to consult the appointed safety representative(s) about the introduction of new equipment or apparatus (see paragraph 47).

Restricting exposure of young people

68   The requirements for risk assessment under regulation 3 of MHSWR[7] and regulation 7 of IRR99 envisage special consideration being given to the employment of young people in work with ionising radiation. The employer has to take into account risks which are a consequence of their lack of experience, the possible absence of awareness of risks or the fact that young people have not yet fully matured. Regulation 19 of MHSWR places particular requirements on employers of young people with regard to matters such as the risk of accidents. Where the employer's control measures are not sufficient to prevent a significant risk to young people they should undertake the work only if:

(a)   it is necessary for their training;

(b)   he or she is supervised by a competent person; and

(c)   the risk is reduced to the lowest level that is reasonably practicable.

69   Under the provisions of IRR99, young people cannot be designated as classified persons (see regulation 20) and are subject to dose limits that are substantially lower than those for adult employees (see paragraph 189).

Application to medical exposures

70   This regulation does not apply to the protection of those undergoing a medical examination or treatment, which is covered by the Ionising Radiation (Protection of Persons Undergoing Medical Examination or Treatment) Regulations 1988[12] (POPUMET) and successor regulations made to implement the Medical Exposure Directive 97/43/Euratom.[4] However, it does apply to staff who carry out those exposures, to members of the public and to 'comforters and carers'.

Protection for the skin of the hand

71   Radioactive materials, including those in the form of sealed sources, should not be held or directly manipulated in the hand (or close to the hand) if it is practicable for the task to be completed by other means, unless the skin of the hand is unlikely to receive a significant dose and the employee is unlikely to become significantly contaminated with radioactive substances.

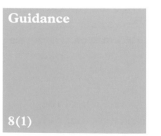

72   This ACOP advice implies that, normally, for research and other laboratories, small quantities of radioactive substances should only be handled where there are no practicable alternatives. In general, it would be practicable to use local shielding and protective gloves in these circumstances so that external radiation and skin contamination risks are controlled effectively.

73   In general, it is impracticable to avoid handling a syringe containing a radiopharmaceutical for the injection of a patient, although it would usually be

practicable to provide a syringe shield to restrict exposure to external radiation. Conversely, it should always be practicable to avoid handling or working close to a sealed source used, for example in industrial radiography.

(2)   Without prejudice to the generality of paragraph (1), a radiation employer shall -

(a)   so far as is reasonably practicable achieve the restriction of exposure to ionising radiation required under that paragraph by means of engineering controls and design features and in addition by the provision and use of safety features and warning devices; and

(b)   in addition to sub-paragraph (a) above, provide such systems of work as will, so far as is reasonably practicable, restrict the exposure to ionising radiation of employees and other persons; and

(c)   in addition to sub-paragraphs (a) and (b) above, where it is reasonably practicable to further restrict exposure to ionising radiation by means of personal protective equipment, provide employees or other persons with adequate and suitable personal protective equipment (including respiratory protective equipment) unless the use of personal protective equipment of a particular kind is not appropriate having regard to the nature of the work or the circumstances of the particular case.

Hierarchy of control measures

74   Regulation 8(2) establishes a hierarchy of control measures for restricting exposure. First and foremost, in any work with ionising radiation, radiation employers should take action to control the doses received by their employees and other people by engineered means. Only after these have been applied should consideration be given to the use of supporting systems of work. Lastly, radiation employers should provide personal protective equipment to further restrict exposure where this is reasonably practicable.

75   Usually, radiation employers will achieve the most effective restriction of exposure by taking control measures at an early stage, for example when the facility or device is being planned and designed. That way, the dose control mechanisms can be incorporated into the construction of the facility. Paragraphs 119-138 give guidance on the use of dose constraints at the design or planning stage of work with ionising radiation.

Types of physical control measures

76   Engineering controls and design features are normally those physical controls built into the facility or device; they include all aspects of the design and construction which restrict exposure. Where appropriate, these controls will be intrinsic to the operation of the facility or device, for example by the construction of suitable containment and shielding of sources and the design of safety-related control systems to ensure that radiation sources are accessible to no greater extent than is necessary for the work being undertaken by the radiation employer. In other cases, engineering controls and design features will be put in place specifically to allow the safe use of the device. Examples include beam collimation, local shielding to reduce emitted radiation and local containment, ventilation or other steps aimed at minimising contamination during work with unsealed radioactive materials.

77   Safety features are intended both to help ensure the safe use of the equipment in normal operation and to prevent unintended exposure in the event of a failure of control devices or systems of work. Examples include

locks on exposure controls, search and lock-up systems, door interlocks for enclosures and emergency exposure controls ('off' buttons).

78 Warning devices indicate the status of the equipment in normal operation and alert operators to faults or failures which have occurred and which reduce the safety integrity of the installation. These devices will not of themselves prevent exposure but they will indicate to the equipment operators what action to take and not to take. Examples include pre-exposure and exposure signals and external radiation or contamination alarms.

Enclosure and radiation shielding

79 Where reasonably practicable, work involving exposure to external radiation should be done in a room, enclosure, cabinet or purpose-made structure which is provided with adequate shielding. In other cases, adequate local shielding should be used as far as reasonably practicable. Shielding, including beam collimation, will normally be adequate if designed to reduce dose rates below 7.5 microsieverts per hour in specific locations where persons will be working. If the device is designed for use in public areas or where there is continuous access to the working area by employees or other persons not directly involved in the work, the shielding should be designed to reduce dose rates to the lowest level that is reasonably practicable. In this case, the dose rate should be so low that it is unnecessary to designate the area around the device as a supervised area.

80 In many cases, shielding will either form part of the equipment (eg covers, shutters and collimators) or an enclosure around the device (eg a room or purpose-made structure). Local shielding around sources - including purpose-made covers, drapes, free-standing screens and even bags of lead shot - can also be used to restrict exposure where an enclosure is not reasonably practicable.

Fluoroscopic devices

81 Fluoroscopic devices should be provided with viewing facilities which do not permit direct vision of the fluoroscopy screen.

82 For medical imaging applications and security inspection, an image intensifier would normally be used with a TV camera and monitor.

Work with unsealed materials

83 Radiation employers should give priority to the containment of radioactive substances as a means of preventing dispersal or contamination. Where such containment alone is not sufficient to give the required protection, ventilation should be provided. A building, room or enclosure being built or modified for work with unsealed radioactive material should incorporate design features which take into account the risk of contamination likely to arise from the work. In particular, radiation employers should take steps to ensure ease of cleaning and decontamination of worktops, floors, etc. There should also be provision for safe decommissioning or dismantling of equipment which may have become internally contaminated.

84 The radiation employer should aim to select, as appropriate, fume cupboards, sterile cabinets, glove boxes, ducting, fan assemblies, filtration units and other components of the ventilation system which have been

designed and constructed specifically for radioactively contaminated atmospheres. Good design and construction will ensure that such equipment is constructed to facilitate maintenance, cleaning, and decontamination; also disposal of filter media and other such contaminated parts of the system should be achieved with minimum personal exposure to radiation. The design and operation of the ventilation system will need to satisfy any discharge authorisation requirements of the Radioactive Substances Act 1993[6] (RSA93) enforced by the relevant environment agency. In addition, employers registered under RSA93 (or covered by an exemption order made under that Act) are required to minimise any waste arising from that work.

85   The provision and cleaning of washing facilities and changing facilities are dealt with under regulation 18(7) (see paragraphs 329-335).

86   Effective use of ventilation is the primary means of reducing levels of radon 222 and radon decay products (which present the main health risk) in the working environment of mines and certain workplaces in radon affected areas.

Exposure controls

87   Where control systems permit, interlocks or trapped key systems should be provided and properly used where they can prevent access to high dose rate enclosures (for example in which employed persons could receive an effective dose greater than 20 millisieverts or an equivalent dose in excess of a dose limit within several minutes when radiation emission is under way). They should be fitted so that the control system will ensure an exposure:

(a)   cannot commence while the access door, access hatch, cover or appropriate barrier to the enclosure is open;

(b)   is interrupted if the access door, access hatch, cover or barrier is opened; and

(c)   does not recommence on the mere act of closing a door, access hatch, cover or barrier.

88   Normally, it should be reasonably practicable to design control units for X-ray generators (and, where appropriate, radioactive source containers) to prevent unintended and accidental exposure. To avoid any possible confusion, control switches usually need to be labelled clearly and unambiguously.

89   The radiation employer will need to consider how an exposure will be terminated or a radioactive source returned to its shielding if the normal control system malfunctions.

90   On many inspection and gauging devices the material being examined is taken past the imaging or sensing system. Routinely, there will be no need for a radiation beam if the material transfer system is not operating. Therefore, it is normally appropriate for the exposure control system to terminate an exposure or close the shielding shutter when the transfer system stops or is stopped.

91   For high dose rate radiation enclosures, the radiation employer may need to take particular steps to make sure that no employee remains inside when exposures commence. Examples of such facilities include: shielded enclosures used for industrial X-ray equipment and linear accelerators, medical radiotherapy suites, gamma and electron beam sterilisation facilities and detector cells on a research accelerator. Facilities in which it may be possible

to exceed a relevant limit on equivalent dose to part of the body (eg to the hand) within a few minutes would include enclosures around X-ray optical equipment.

92   The ACOP advice in paragraph 87 means that effective interlock devices should normally be designed and installed in such a manner that if they fail to operate correctly no exposure of people can occur. Although it should be reasonably practicable to install these devices where enclosures or cabinets are provided around X-ray generators, it may sometimes be more difficult to provide them for sealed source equipment because of the method of controlling the source. However, the installation of such devices with this type of equipment is often reasonably practicable.

93   Robust systems of work should normally prevent the possibility of people being trapped in a high dose rate enclosure (see paragraphs 104-107). The consequences of a person being trapped in such an enclosure are potentially very serious. An assessment under regulation 7 may indicate a need for further action, for example it might be appropriate to place emergency devices such as 'off' buttons in appropriate locations to allow a person to prevent or quickly interrupt the emission of ionising radiation.

94   Where there is a risk of receiving an exposure within a few minutes that could cause deterministic effects (eg erythema) it would normally be reasonably practicable for employers to install engineered search and lock-up systems into the exposure control system. These systems require the full area of the enclosure to be properly checked and vacated before the exposure from X-ray equipment or an accelerator can be initiated. The check would be confirmed by the operation of 'search' buttons within a predetermined time before the area is closed and irradiation begins. Search and lock-up systems may be appropriate for facilities where very intense sources are used, such as industrial linear accelerator enclosures, gamma and electron beam sterilisation facilities and detector cells on a research accelerator.

95   If, despite the use of exposure controls, employees might become trapped in a high radiation dose enclosure, for example an industrial radiography enclosure, it will normally be appropriate to install a suitable alarm system to make them aware that an exposure is imminent. In such a situation, employees will need access to an 'emergency' control button which will enable them to prevent the initiation of the exposure or to terminate the exposure themselves. Otherwise, they should be able to alert the operator(s) outside the enclosure to their presence and to take refuge in a shielded part of the enclosure without passing through the main beam. The dose rate in that shielded refuge should preferably be below 7.5 microsieverts per hour during any exposure.

Key-operated safety devices

96   Where there is a risk of significant exposure arising from unauthorised or malicious operation of X-ray generators or radioactive source containers, radiation employers should make use of equipment which has been fitted with locking-off arrangements to prevent its uncontrolled use.

97   The initiation of exposures should be under key control, or by some equally effective means, so as to prevent unintended or accidental emission of a radiation beam or exposure of a source. This is particularly important where the control point is remote from the equipment which will be activated or there is general access to equipment by members of the public or personnel who are not undertaking the work with ionising radiation.

98   Where key-operated devices are provided, arrangements will be needed to ensure that keys are only available to authorised employees. Trapped key systems may help to prevent unauthorised or accidental operation of the equipment.

Warning devices

99   Sources of ionising radiation which can give rise to significant exposure in a very short time should be fitted with suitable warning devices which:

(a)   indicate for a radioactive source whether it is in or out of its shielding (or the exposure shutter is open or closed);

(b)   indicate for an X-ray generator when the tube is in a state of readiness to emit radiation and, except for diagnostic radiology, give a signal when the useful beam is about to be emitted and a distinguishable signal when the emission is under way unless this is impracticable;

(c)   for X-ray generators other than those used for diagnostic radiology, are designed to be automatic and fail-safe, ie if the warning device itself fails the exposure will not proceed.

100   The radiation employer should make sure that warning signals can be seen or heard by all those people who need to know the status of the radiation equipment for protection purposes.

101   Automatic warning devices should be reasonably practicable for most X-ray generators and some sealed sources. Therefore, radiation employers using large sealed sources will need to consider how signals could be provided. In particular, automatic generation of the pre-exposure and exposure warning signals is possible where there is a fully automated mechanism for the exposure and retraction of the source. If it is not reasonably practicable to follow the ACOP advice in sub-paragraph 99(c) that warning devices should normally be fail-safe, daily checks that such devices are working might suffice in the short term, at least. Where a visual device relies on a single bulb, it may be sufficient to modify the device to use two bulbs.

102   In some cases, warning signals will need to be seen or heard by people who may be outside the immediate area. Therefore, the radiation employer might provide explanatory notices for people not directly involved in the work, pointing out any action they should take in response to a warning signal. These notices will be of little use if they fail to explain clearly the significance of the different signals and the action to be taken. The radiation employer might need to consult the appointed safety representative(s) for the area about such notices.

Manual warning signals, radiation monitors and warning labels

103   Where equipment containing a radioactive source is not fitted with automatic warning signals, it may be appropriate for the operator to make use of a manual system to generate clear and distinguishable pre-exposure and exposure warning signals. Signals are particularly important when people in or around the designated area need to be aware of the status of the source. Notices and signs can explain the significance of these signals and the action to be taken.

Systems of work

104 The engineered safety features should, so far as reasonably practicable, be supported by systems of work to be followed by employees and other people when present in the vicinity of radiation equipment. In circumstances where a person working with ionising radiation could receive an overexposure in a relatively short period of time, for example several minutes or less, these systems of work should normally be formalised. This may involve the issue of a certificate of entry specifying the detailed working arrangement for that entry. These formalised systems are frequently known as 'permit-to-work systems' and they allow strict management control over the conditions in which work will proceed, how it will be done, and how it will be supervised.

105 The employer should normally consult the appointed safety representative(s) relevant to the area about any proposal to introduce a permit-to-work system.

106 In many situations the radiation employer can take simple steps to control the exposure of people, for example by ensuring that:

(a) the number of people present in high dose rate or contaminated areas is limited and the time they remain there minimised;

(b) in interventional radiology, staff not required to be close to the couch and involved in the direct clinical examination or care of the patient remain away from the high dose rate areas (which requires a good understanding of dose rate contours around the couch) and preferably behind a shielded screen;

(c) in a storeroom containing a large number of different radioactive sources, any areas or zones in which the dose rate is significantly high are well known and work in the storeroom is organised so that the time spent in these areas or zones is minimised, taking account of the impact on external dose rates outside the storeroom;

(d) for transport of radioactive material by road, the system of loading is such that packages emitting a higher dose rate are positioned furthest from the driver's seat, thus reducing the driver's exposure;

(e) work with unsealed radioactive substances is carried out in a fume cupboard and lipped trays with absorbent lining are used to contain minor spillages.

107 Where appropriate, any signs used to indicate high dose rate areas should conform to the Health and Safety (Safety Signs and Signals) Regulations 1996.[13] Where the area concerned has been designated as a controlled area or a supervised area, the requirements of regulation 18 may also be relevant.

108 It is crucially important that detailed safe systems of work are produced for work on equipment in which any of the engineered safety systems have been disabled. Close supervision of the work may be necessary. Examples include:

(a) the loading or unloading of sealed sources in or from equipment where normal safeguards for restricting exposure cannot be followed;

(b) the setting up and alignment of X-ray optics equipment; and

(c) maintenance work on a thickness gauge incorporating a sealed source which involves any work inside the gauge head or with the gauge heads split.

Guidance

A permit-to-work (see paragraph 104) may be necessary in certain cases.

109   Where appropriate, the radiation employer should consider providing systems for cleaning up minor spillage during normal work, to prevent the build-up of radioactive contamination. If it is not reasonably practicable to clean up all the contamination in an area or on a piece of work equipment, special restrictions might then need to be applied to future work in that area. If the contamination consists of short-lived radionuclides, it may be prudent to allow these to decay rather than to take immediate action to clean up the area. There will be circumstances, however, where it is necessary to clean up radioactive contamination even at the expense of increasing the exposure of a group of employees involved in the clean up, especially where the public have access to that area.

110   Effective training for employees (regulation 14) will enable them to follow systems of work and to interpret and respond to any warning signals or supporting notices which they may encounter in the course of their work.

ACOP

8(2)

Monitoring radiation levels as part of a system of work

111   The radiation employer should require a check to be made with a suitable radiation monitoring instrument after each exposure using high dose rate sealed source equipment (such as that generally used for industrial radiography or processing of products) unless reliance can be placed on effective devices to ensure that the equipment has been restored to a safe state. The purpose is to establish that the sealed source has fully retracted to its shielded position and that the area is safe to enter.

Guidance

112   Radiation employers may need to ensure that monitoring instruments are used as part of the system of work for certain non-routine operations such as maintenance work on high dose rate sealed source equipment, as well as providing such instruments for routine industrial radiography work or processing of products see also regulation 19.

Adequate support for personnel

113   All employers will need to take decisions on the number of trained personnel required to be present to allow a job of work to proceed safely. For example, in site radiography work the radiographer will normally require the support of at least one other person. This assistant will usually need to patrol the boundaries of the controlled area to ensure that access is restricted (see regulation 18) and perform the necessary radiation monitoring of that area (see regulation 19) and would be available to assist with implementing the contingency plan (see regulation 12).

ACOP

8(2)

Personal protective equipment

114   The term 'adequate' in regulation 8(2)(c) refers to the ability of the equipment to protect the wearer. The term 'suitable' refers to the correct matching of the equipment to the job and the person. To be considered 'adequate and suitable' personal protective equipment should be correctly selected and used.

Guidance

8(2)

115   Personal protective equipment (PPE) includes respiratory protective equipment (RPE), protective clothing, footwear and equipment to protect the eyes. Types of PPE specific to protection against external ionising radiation include lead aprons and gloves. Various types of respiratory protective

equipment (including pressurised suits) provide protection specific to the risk of inhaling radioactive material.

116 The risk assessment should be used to decide on the choice of PPE (see paragraph 45(c)). The purpose of the assessment, in this case, is to ensure that the employer chooses PPE which is adequate and suitable, ie correct for the circumstances of use. For RPE this implies that it provides an adequate margin of safety and is matched to the job, the environment, the anticipated air concentration of radioactive material and to the wearer. The performance of RPE which relies on a tight-fitting facepiece will be adversely affected if there is not good contact between the wearer's skin and the face seal of the mask, eg because the wearer is not clean shaven. RPE from more than one manufacturer and of more than one type may be needed to meet the face fit requirements of all the employees in a particular area. Proper testing of the correct fit for facepieces intended to fit tightly is a necessary part of the selection process.

117 Sometimes, PPE could render the wearer liable to other forms of risk greater than that arising from the ionising radiation. In these circumstances, it might not be appropriate to use PPE. Examples might include use of bulky breathing apparatus in areas with limited space for free movement, or the wearing of a lead apron by an employee who is unable to carry its weight without risk of injury.

118 HSE guidance on the Personal Protective Equipment at Work Regulations 1992[14] provides much advice which is equally applicable to protection against ionising radiation. It offers advice about the selection and use of the types of PPE. HSE publication HSG53[15] gives further advice on the selection, use and maintenance of suitable RPE; Annex 1 of HSG53 gives special guidance relating to radioactive substances.

Dose constraints

(3)   Where it is appropriate to do so at the planning stage of radiation protection, dose constraints shall be used in restricting exposure to ionising radiation pursuant to paragraph (1).

What are dose constraints?

119 A dose constraint is an upper level of individual dose specified by the employer for use at the design or planning stage. It is one of many tools for helping to restrict individual exposures as far as reasonably practicable. Dose constraints may be used to consider the best plan or design for an individual task or event or the introduction of a new facility. However, they are not intended to be used as investigation levels once a decision has been taken about the most appropriate design or plan (see paragraphs 152-156).

120 In general, the value assigned to a dose constraint is intended to represent a level of dose (or some other measurable quantity) which ought to be achieved in a well-managed practice.

121 Realistic predictions of individual doses associated with any proposed control measures for restricting exposure would be compared with the value selected for the dose constraint. If the predicted doses exceeded the value of the relevant dose constraint, the radiation employer would normally be expected to choose better control measures. These should then lead to predicted doses below the dose constraint. Therefore, a dose constraint should

help to filter out options for radiation protection that could lead to unreasonably high levels of individual dose, even though the collective dose for the workforce as a whole is optimised.

122 In some cases, the radiation employer may decide that it is acceptable for predicted doses to exceed the dose constraint where, for example, other health and safety risks have to be taken into account in selecting the most appropriate control measures.

When is it appropriate to use dose constraints?

123 Dose constraints are appropriate for 'comforters and carers' (see paragraph 127) because these individuals are not subject to dose limits. Also, dose constraints are generally appropriate for members of the public who may be affected by direct radiation or contamination arising from work activities, not least because those individuals may be exposed to more than one such source. However, for occupational exposure, dose constraints may only be appropriate in a limited number of situations.

124 The need to establish dose constraints is best determined as part of the risk assessment process (see paragraph 45). Where the employer establishes one or more dose constraints it would be sensible to record these as part of the arrangements made under the written safety policy required by HSW Act.[2]

125 Dose constraints are meant to represent levels which are normally obtainable in the particular type of work activity in well-managed operations using effective physical controls and systems of work to restrict exposure. Where an employer decides it is appropriate to establish dose constraints for particular types of work, these could be based on past operating experience and on any recommendations from relevant professional bodies or trade associations.

8(3)

Dose constraints for comforters and carers

126 It should always be appropriate to use dose constraints in restricting exposure for comforters and carers.

127 'Comforters and carers' are defined as 'individuals who (other than as part of their occupation) knowingly and willingly incur an exposure to ionising radiation in the support and comfort of another person who is undergoing, or who has undergone, a medical exposure'. That is, they will normally be relatives or friends of the patient; they are not health-care employees.

128 They may include members of the public who, for example:

(a) visit patients in hospital after those patients have been administered with radiopharmaceuticals (most notably for therapeutic purposes) or have undergone brachytherapy;

(b) offer support for those patients at home after they have been discharged from hospital; or

(c) (in some cases) offer support to a young child or disabled person while that child or person receives a diagnostic X-ray examination;

and are likely to receive 1 millisievert or more in a year resulting from direct radiation or contamination during the comfort and support they offer.

129 Radiation employers will need to make suitable arrangements to satisfy themselves that these individuals are aware of the risks involved in supporting and comforting a patient and are willing to incur the exposures they will receive. These arrangements may involve effective communication with the patient or directly with the comforter and carer.

130 The exposure of comforters and carers should normally be controlled, as far as reasonably practicable, by using time, distance and shielding, taking into account the wishes of the individual to offer comfort and support to the patient. As comforters and carers are not subject to dose limits, the dose constraint is important as a means of helping to plan general arrangements for restricting any unnecessary exposure of such people.

131 In most cases, comforters and carers would normally be expected to keep to these general arrangements. However, they may choose to depart from them, for example by spending more time with a seriously ill patient than is recommended, thereby incurring a dose greater than the numerical value of the dose constraint. That is perfectly reasonable provided that they do so willingly and are aware that they may incur a small additional risk from this increased exposure.

132 The National Radiological Protection Board has recommended that in this context, medical exposures incurred by individuals, other than as part of their own medical diagnosis or treatment, should not in general exceed 5 millisieverts from their involvement in one series or course of treatment. This value was selected mainly on the basis of experience. Normally it should be possible to design procedures that will keep doses received by comforters and carers below this level while they offer comfort and support to patients. Radiation employers in the health-care sector considering arrangements to protect comforters and carers may wish to take account of this recommended value (together with advice from professional bodies) in selecting appropriate dose constraints.

133 Health-care employers would normally take account of any dose constraints in devising standard procedures for restricting the exposure of comforters and carers visiting patients administered with radiopharmaceuticals and for deciding when such patients ought to be discharged. In practice, these arrangements will often reflect general good practice for restricting the exposure of people who act as comforters and carers.

Dose constraints for members of the public

134 Where the radiation employer anticipates that the new work activity or facility is likely to expose members of the public to direct radiation or contamination, it may be appropriate to use a dose constraint. The National Radiological Protection Board has recommended that the constraint on optimisation for a single new source should not exceed 0.3 millisieverts a year. Radiation employers may wish to take this recommendation into account in establishing a dose constraint for members of the public. The constraint would normally be applied to estimates of dose for members of the critical group likely to receive the highest average dose from the work.

135 Radiation employers in the health-care sector may need to consider the use of dose constraints for members of the public who are not comforters and carers but are likely to receive some exposure as a result of sharing transport or accommodation with a patient who has received a therapeutic administration of a radiopharmaceutical (see also paragraphs 196 and 197). In practice, however, the radiation employer may decide that sufficient action has been taken by following accepted practice on the release of such patients from hospital.

38

Dose constraints for occupational exposure

136 Dose constraints for occupational exposure are only likely to be appropriate where individual doses from a single type of radiation source will be a significant fraction of the dose limit (ie at the rate of a few millisieverts a year). Dose constraints are not likely to be appropriate for occupational exposures resulting from the following types of work with ionising radiation:

(a) diagnostic radiology, nuclear medicine, most radiotherapy and other medical exposures;

(b) most work in the non-nuclear industrial sectors; and

(c) teaching and most research activities;

where employee doses tend to be low. The main exception would be for special types of work (eg some interventional radiology), where effective doses are likely to be more than a few millisieverts a year.

137 Even in specialised areas, such as industrial radiography, where individual doses are sometimes relatively high, it may not be appropriate to establish dose constraints for planning individual jobs unless adequate dose information is available for that type of work. If the radiation employer has such information, it should be feasible to choose a dose constraint which is representative of a well-managed operation, for example radiography of steam tubes during a conventional power station outage.

138 The nuclear sector has considerable experience in developing dose databases relating to good practice. Therefore, dose constraints may be appropriate for some well-defined operations in this sector. Any dose constraints established by such employers should be consistent with HSE's Safety Assessment Principles.[16]

Proper use of systems of work and PPE

(4)   An employer who provides any system of work or personal protective equipment pursuant to this regulation shall take all reasonable steps to ensure that it is properly used or applied as the case may be.

139 Procedures should be established by the radiation employer to ensure the proper use of control measures such as systems of work and items of PPE provided to comply with regulation 9(2). They might typically include:

(a) visual checks at appropriate intervals to ensure that these control measures are being properly used; and

(b) prompt remedial action where the control measures break down.

Pregnant and breast-feeding employees

(5)   Without prejudice to paragraph (1), a radiation employer shall ensure, that -

(a)   in relation to an employee who is pregnant, the conditions of exposure are such that, after her employer has been notified of the pregnancy, the equivalent dose to the foetus is unlikely to exceed 1 mSv during the remainder of the pregnancy; and

(b)   in relation to an employee who is breastfeeding, the conditions of exposure are restricted so as to prevent significant bodily contamination of that employee.

Assessing risks for female employees

140   The assessment required by regulations 3 and 16 of MHSWR[7] should take account of risks to the health and safety of a new or expectant mother at work, or to that of her baby or foetus. For any new activities, the prior risk assessment required by regulation 7 of IRR99 should indicate to the radiation employer what exposures female employees are likely to receive in particular working areas. The ACOP at paragraph 217 advises that the RPA should normally be involved in that assessment. The risk assessment will help the employer to focus attention on those areas where a pregnant employee might be exposed to such an extent that her foetus would receive a dose greater than 1 millisievert during a declared period of pregnancy.

141   For exposure to external radiation, this dose restriction is broadly equivalent to a dose to the surface of the abdomen of a pregnant woman of about 2 millisieverts in many working situations; for example exposure arising from the diagnostic use of X-rays in hospitals. Where exposure is to high-energy radiation, however, the radiation employer will need advice from the RPA about an appropriate dose restriction. If the woman is likely to receive significant intakes of radionuclides as a result of working conditions in an area (typically where the committed effective dose is of the order of 1 millisievert a year or more), particular care may be needed. In these cases the dose to the foetus may approach or exceed 1 millisievert for certain radionuclides which are preferentially taken up by the tissues of the placenta and foetus.

142   Employers will also need to take into account risks relating to female employees who are breast-feeding their infant. The employer's risk assessment should reveal whether there are any working areas where an employee may receive significant bodily contamination which could pose a risk to her or her infant. The employer can take account of any control measures provided to prevent such contamination, including the use of PPE, provided that effective steps have been taken to ensure it is used at all times. In deciding whether bodily contamination may be significant, employers should be aware that certain radionuclides are likely to become concentrated in breast milk and that the dose to the infant may be of much greater concern than the dose to the woman.

Declaration of pregnancy

143   Implementation of the control measures set out in this regulation depends on the employee informing the employer of her condition. It follows that female employees need to understand the importance of notifying the employer in writing, even if they would prefer to keep their condition completely confidential. Regulation 14 requires employers of female employees to ensure that they are informed about the possible risks and the importance of informing the employer in writing as soon as they are aware of their pregnancy. Employers may wish to make arrangements for female employees to receive counselling from the appointed doctor or an occupational health service to discuss the possible risks involved. These arrangements might also provide an opportunity for a female employee who becomes pregnant to discuss her condition with a health professional and to alert the employer to her condition accordingly.

144   The employer may request confirmation of the pregnancy by means of a certificate from a registered medical practitioner or a registered midwife or otherwise in writing. If this certificate has not been produced within a reasonable period of time, the employer is not bound to maintain any changes to the working conditions or other arrangements made for the woman under this regulation and regulation 16 of MHSWR[7] (see paragraph 149).

145 If the employee informs her employer that she is pregnant for the purpose of any other statutory requirements, such as maternity pay, this will be sufficient in the absence of any earlier notification, and the employer should act on the information for the purposes of this regulation.

Return from maternity leave

146 Where an employee returns from maternity leave to work involving unsealed sources and the radiation employer's assessment shows that bodily contamination is reasonably foreseeable, it is normally advisable to assume that she is breastfeeding and to take appropriate action, notwithstanding the provision in regulation 8(6). The employer's action might include giving advice about the possible risks (regulation 14) and altering her working conditions. The appointed doctor or occupational health service may provide counselling, if appropriate. If the employee continues to breast-feed her infant for more than six months after the birth it would be sensible for her to notify her employer to ensure that any special measures to prevent bodily contamination are maintained.

Need for co-operation between employers

147 Where the radiation employer engages a contractor to undertake any work in which that contractor's employees are likely to be exposed to ionising radiation, the radiation employer should make arrangements with that employer (under regulation 15) to be informed if any of the contractor's employees:

(a) have declared themselves to be pregnant; or

(b) in cases where bodily contamination is reasonably foreseeable, they are breastfeeding;

unless the woman, her foetus or her infant would not be at risk from ionising radiation during that work. Similar considerations apply to self-employed female contractors.

Action required by the radiation employer

148 After being informed of the woman's condition, the radiation employer should decide if any restrictions are necessary in the particular case, taking account of the results of the risk assessment.

149 If the assessment shows some action is necessary, this may be limited to avoiding work:

(a) with large sources of external radiation where there is a reasonably foreseeable risk that pregnant employees may receive a significant accidental exposure; or

(b) where there is a significant risk from intakes of radionuclides.

150 If such risks cannot be avoided in line with regulation 8(5) of IRR99, regulation 16 of MHSWR[7] requires the employer to:

(a) alter the hours of work of the individual or, where this would not be reasonable;

(b) identify and offer her suitable alternative work that is available; or where this is not feasible;

(c) suspend her from work on full pay.

41

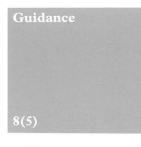
The Employment Protection (Consolidation) Act 1978[17] and the Employment Rights Act 1996[18] (which is the responsibility of the Department of Trade and Industry (DTI)) require that this suspension should be on full pay. Employment rights are enforced through the employment tribunals.

151 Further guidance on new and expectant mothers is contained in the HSE publication HSG122.[19] The booklets PL705[20] and PL958,[21] which both cover the maternity suspension provisions, are available from DTI.

(6)   Nothing in paragraph (5) shall require the radiation employer to take any action in relation to an employee until she has notified her employer in writing that she is pregnant or breastfeeding and the radiation employer has been made aware, or should reasonably have been expected to be aware, of that fact.

Formal investigation levels

(7)   Every employer shall, for the purpose of determining whether the requirements of paragraph (1) are being met, ensure that an investigation is carried out forthwith when the effective dose of ionising radiation received by any of his employees for the first time in any calendar year exceeds 15 mSv or such other lower effective dose as the employer may specify, which dose shall be specified in writing in local rules made pursuant to regulation 17(1) or, where local rules are not required, by other suitable means.

152 The purpose of this provision is to trigger a review of working conditions when an employee (or a group of employees undertaking similar work) have recorded doses which exceed the specified investigation level, to make sure that exposure is being restricted as far as is reasonably practicable. The investigation has a different purpose from that required by regulation 25 concerning possible overexposures.

153 The duty to carry out an investigation is placed on the actual employer of the person whose recorded dose has exceeded the investigation level. In most cases, this will probably be an employer who is working with ionising radiation (ie radiation employer). However, the employer might be a contractor (eg a scaffolding contractor or a cleaning company) working on various sites occupied by radiation employers. The investigation may have to take account of work with ionising radiation undertaken at all these different sites throughout the calendar year.

154 The regulation encourages employers to specify an appropriate investigation level for relevant groups of employees which will often be significantly less than 15 millisieverts a year. Normally, the employer would take into account the profile of doses for employees as a whole (or particular groups) and seek advice from the RPA before selecting an appropriate investigation level. The appointed safety representative(s) or the established safety committee may be consulted about the choice of the investigation level. The default level is 15 millisieverts a year; employers are not allowed to specify an investigation level higher than that.

155 The employer could select different investigation levels for different sites or different groups of employees, where appropriate.

156 Employers may wish to have arrangements for reviewing any unusually high doses reported in dose summaries for classified persons by an approved dosimetry service (see regulation 21(3)(b)) or for other people entering controlled areas (see regulation 18(5)). Such arrangements will provide an early warning if an employee's cumulative dose for the year is approaching the investigation level. In these cases, the employer may decide it is appropriate to take further measures to restrict exposure before a formal investigation becomes necessary.

157 A formal investigation under regulation 8(7) should be aimed at checking the adequacy of the control measures provided to restrict exposure, to see if further engineering controls, etc are appropriate. It might include such matters as:

(a) details of the work routine of the individual and immediate work colleagues for the year so far;

(b) evidence about the involvement of the individual in any known incidents in which they may have received an unusual exposure;

(c) details of that person's estimated exposures by task or relevant dose assessment period, compared with estimated exposures of work colleagues undertaking similar work;

(d) results of any special radiation survey in the areas where the person worked, to identify any deterioration in physical control measures; and

(e) evidence from the radiation protection supervisor (RPS), the individual concerned and work colleagues about adherence to local rules or deficiencies in those rules in the light of changes to work practices.

158 An effective investigation will normally produce firm conclusions about the need for further control measures (or the better application of current controls). Those conclusions would normally be recorded together with the main findings in an investigation report (see paragraph 162). If those conducting the investigation conclude that further action is appropriate they should consider making recommendations to senior management, as appropriate. That action might include changes to physical control measures or systems of work and planning exposures for the individual or group for the remainder of the year.

159 In accordance with the ACOP advice in paragraph 217, the employer should normally consult the RPA about such investigations.

160 Where groups of employees are engaged in essentially similar work in the same type of environment, only one investigation may be needed if two or more individuals receive doses above the investigation level.

161 The employer would normally consult any appointed safety representative about the formal investigation.

162 The employer will need some means of demonstrating to an HSE inspector or others that an adequate investigation has been carried out. Retaining a copy of the investigation report for at least two years should be sufficient for this purpose.

Regulation 9

Personal protective equipment

(1)   Any personal protective equipment provided by an employer pursuant to regulation 8 shall comply with any provision in the Personal Protective Equipment (EC Directive) Regulations 1992[a] which is applicable to that item of personal protective equipment.

(2) Where in the case of respiratory protective equipment no provision of the Regulations referred to in paragraph (1) applies, that respiratory protective equipment shall satisfy the requirements of regulation 8 only if it is of a type, or conforms to a standard, approved in either case by the Executive.

(a) SI 1992/3139; as amended by SI 1993/3074, SI 1994/2326 and SI 1996/3039.

163 Requirements governing the 'CE' marking, design, manufacture and marketing of PPE derive from a European Directive (89/686/EEC)[22] which is implemented in the UK by the Personal Protective Equipment (EC Directive) Regulations 1992.[23] Requirements governing the use of PPE derive from a European Directive (89/656/EEC)[24] which applies to protection against most hazardous situations and agents, including ionising radiation. It was mainly implemented in Great Britain by the Personal Protective Equipment at Work Regulations 1992,[14] except for requirements specifically relating to ionising radiation. In general, the employer should select CE-marked equipment in order to satisfy regulation 9(1). HSE approval of RPE ceased on 30 June 1995 but employers can continue to use HSE-approved equipment made before 1 July 1995, as long as it is suitable and maintained to perform correctly. Further advice is given in HSE guidance on those Regulations and in the HSE publication HSG53.[15]

164 The employer should consider the need for RPA advice on the selection of adequate and suitable PPE for restricting exposure. The appointed safety representative(s) might also be consulted about the suitability of PPE for the wearer.

(3) Every radiation employer shall ensure that appropriate accommodation is provided for personal protective equipment when it is not being worn.

Maintenance and examination of engineering controls etc and personal protective equipment

(1) A radiation employer who provides any engineering control, design feature, safety feature or warning device to meet the requirements of regulation 8(2)(a) shall ensure -

(a) that any such control, feature or device is properly maintained; and

(b) where appropriate, that thorough examinations and tests of such controls, features or devices are carried out at suitable intervals.

165 All active engineering controls and design features (eg local exhaust ventilation systems), safety features (eg electromechanical interlocks) and warning devices should be subjected to a regime of examination and test at suitable intervals.

General advice on maintenance of controls

166 'Maintained' is defined under regulation 2 in relation to plant, apparatus, equipment and facilities. 'Maintenance' means 'maintained in an efficient state, in efficient working order and good repair'.

167 The objective of this regulation is to ensure that all physical control measures such as shielding, enclosure, ventilation, safety features and warning devices provided in accordance with regulation 8(2), perform as originally intended. It may not be reasonably practicable to ensure at all times that

complex enclosures or safety devices are in an efficient state and in efficient working order. In such cases, where failures are revealed, eg during the examinations and tests, the radiation employer should either cease that work with ionising radiation or take other effective action under regulation 8(2) to restrict exposure until the enclosure or device is restored to efficient working order.

168   Where the radiation employer holds a site licence granted under the Nuclear Installations Act 1965[5] and controls, features or devices are already maintained and tested in accordance with relevant conditions in that licence, such actions should be sufficient to comply with the requirements of this regulation.

Nature of inspections and tests

169   The radiation employer will need to institute formal programmes for the inspection and test of active design features, such as exhaust ventilation systems. However, for passive engineering controls and design features, the employer will normally use data from monitoring programmes provided under regulation 19 to check the effectiveness of radiation shielding and the systems in place for containment and minimisation of contamination for work with unsealed materials. For example, routine contamination monitoring can be used to check the continuing suitability of the easy-to-clean surface on a table in a laboratory. In this case, the radiation employer should make use of the monitoring information to check that the surface finish is maintained to the required standard.

170   The effectiveness and performance of the active engineering controls would normally be assured and checked when any acceptance test is done, before the equipment is first put into use (see paragraph 67). The tests and inspections done under this regulation can be seen as repeating a sample of these original acceptance tests at suitable intervals, allowing the employer to maintain the performance of these engineering controls at the initial level.

171   It is advisable to carry out a visual check of all active control measures used to prevent or restrict exposure, provided such checks do not present undue risk to employees undertaking maintenance work. These checks, where appropriate, will probably be needed at frequent intervals, for example perhaps during every shift for safety-critical features.

172   The radiation employer should be satisfied that the person undertaking the examination or test has appropriate competence in order to ensure that the examination or test is sufficiently thorough.

Frequency of tests and inspections

173   The ACOP at paragraph 165 advises that, in addition to the visual checks already mentioned, active control measures such as exhaust ventilation systems, electromechanical interlocks and warning devices will usually need to be examined and tested at suitable intervals. Typically, a suitable interval might be about once a year, depending on the circumstances of the work. If failure of control measures could result in an acute exposure above the investigation level established under regulation 8(7), the frequency of examination and test of that control measure may need to be greater than once a year. In deciding on the required test and inspection frequency, the radiation employer might consider:

(a)   the possible dose implications of a failure;

(b)   the reliability of the control measure; and

(c)   the doses likely to be received by staff while carrying out any examination or test.

Consultation on procedures for maintenance

174   Regulation 13 and Schedule 5 require a radiation employer to consult a suitable RPA on the periodic examination and testing of engineering controls etc appropriate to the equipment they use (see paragraphs 226-227). RPA advice should help the employer to develop a programme of examination and test for appropriate control measures. The appointed safety representative(s) or established safety committee might also be consulted. The employer should also consider the recommendations of the manufacturer or supplier on the programme for the inspection and test of the different control measures.

Recording the results

175   **Sufficient records should be kept of these examinations and tests to enable the radiation employer to identify which controls, features or devices have been examined or tested, what action is required to maintain them and when the next examination or test is due.**

176   The radiation employer would normally retain details such as the date and nature of the test and the name of the person carrying it out, as well as the information specified in the Approved Code of Practice. The details could be held in paper or electronic records. The records might, for example, take the form of a general maintenance log for all controls, features and devices provided under regulation 8(2)(a).

Maintenance etc of personal protective equipment

(2)   Every radiation employer shall ensure that all personal protective equipment provided pursuant to regulation 8 is, where appropriate, thoroughly examined at suitable intervals and is properly maintained and that, in the case of respiratory protective equipment, a suitable record of that examination is made and kept for at least two years from the date on which the examination was made and that the record includes a statement of the condition of the equipment at the time of the examination.

177   The purpose of a thorough examination is to establish that the item being examined is fit for the intended use and that physical deterioration has not occurred since its last thorough examination. Generally, it will not be appropriate to subject disposable PPE such as gloves to thorough examination and test unless they are reused after a working session. Checks for possible radioactive contamination, where appropriate, would usually be necessary on a more frequent basis (eg after each shift or wear period) to comply with the general duty under regulation 8(2)(c).

178   The interval required between thorough examinations will depend on the type of equipment, the hazard against which it is required to protect, the conditions under which it is used, the likelihood of deterioration, and the frequency of use. Thorough examination and testing will need to be carried out by people who are competent to do this work and in accordance with the manufacturer's instructions.

179 In general, for respiratory protective equipment the interval between each examination and test would not normally exceed one month. For self-contained breathing apparatus the examinations should follow the instructions of the manufacturer including, for example, a check on the condition of the air supply. For emergency respiratory protective equipment and ready-to-use PPE kept in sealed packs as recommended by the manufacturer, it should normally be sufficient to see that no obvious deterioration has taken place and it is still 'in date' for use. HSE publication HSG53[15] gives further advice.

Records of examination for personal protective equipment

180 Formal records of examination are only required in respect of respiratory protective equipment. They should contain the means of identification and condition of the equipment, the date of examination and the signature of the person who carried it out. Where large numbers of similar items of respiratory protective equipment are involved, the employer may wish to adopt a dating system. Such a system would be sufficient, provided that it ensures the date of the last examination is known, or can be worked out, and that defective equipment can be identified for removal from service.

Regulation 11

Dose limitation

(1)   Subject to paragraph (2) and to paragraph 5 of Schedule 4, every employer shall ensure that his employees and other persons who are within a class specified in Schedule 4 are not exposed to ionising radiation to an extent that any dose limit specified in Part I of that Schedule for such class of person is exceeded in any calendar year.

181 Assessments of effective dose and equivalent dose from external radiation for the purpose of comparison with the dose limits specified in Schedule 4 of the Regulations should be made using the values and relationships in Annex II of Council Directive 96/29/Euratom.[3]

182 Assessments of committed effective dose and committed equivalent dose following intakes of radionuclides into the body should take account of the dose likely to accrue over a period of 50 years following the intake (up to age 70 for children) and should be attributed to the calendar year of the intake for the purpose of comparison with dose limits.

183 For the assessment of compliance with the dose limits relating to members of the public, realistic estimates should be made of the average effective dose (and where relevant equivalent dose) to representative members of the appropriate reference group for the expected pathways of exposure.

Responsibility for ensuring compliance with dose limits

184 The main requirement in these Regulations is that in regulation 8. The radiation employer must ensure that exposures arising from the work are kept as low as reasonably practicable. Dose limits represent a backstop to that requirement. In most cases, during routine exposure it is unlikely that an individual's dose will approach a relevant dose limit; and doses received by employees who are not normally exposed to ionising radiation in the course of their work should be well below dose limits (see ACOP advice in paragraph 60). However, all employers, whether or not they are radiation employers, must make sure that the cumulative exposure of their employees over the year does not exceed a relevant dose limit (see regulation 15 concerning co-operation between employers).

Demonstrating compliance with dose limits

185 The dose limit quantities 'effective dose' and 'equivalent dose' are quantities specified by the International Commission on Radiological Protection. Effective dose relates to the whole body; equivalent dose relates to the dose received by a tissue or part of the body, for example the skin or the eye. Appendix 1 reproduces the relevant values and relationships in Council Directive 96/29/Euratom[3] which define these dose quantities and the operational quantities to be used for individual monitoring in relation to external radiation.

186 As provided by regulation 2(6), any reference to 'effective dose' means the sum of the effective dose from external radiation and the committed effective dose from internal radiation, unless the context requires otherwise. 'Equivalent dose' also includes the committed equivalent dose from internal radiation. The 'committed dose' from an intake of a radionuclide is the dose assessed to be received by a tissue or organ over at least a 50-year period following the intake (70-year period for children). The intake might occur from inhalation or ingestion of radioactive contamination or from accidental incorporation of radionuclides in wounds such as a cut to the hand.

187 If the biological or physical half-life of a radionuclide is short, all the dose will probably be received in the calendar year in which the intake was received. However, if the radionuclide is long-lived, and is likely to remain in the body for many months or years, the dose is likely to be received over a number of calendar years. By internationally accepted convention, the 'committed dose' for an employee is assigned to an individual's dose record for the calendar year of the intake.

188 Approved dosimetry services (and often RPAs) have the relevant expertise in dealing with these quantities. Dosimetry services approved to assess doses received by classified persons from external radiation are currently expected, as a condition of approval, to use the operational dose quantity $H_p(d)$ specified in Council Directive 96/29/Euratom[3] for personal dose monitoring (see Appendix 1). Those dosimetry services approved to assess committed effective dose (and committed equivalent dose) for classified persons following intakes of radionuclides into the body will generally use the dose coefficients in Annex III of Council Directive 96/29/Euratom, where appropriate.

Application of dose limits to different classes of person

189 Schedule 4, Part I has the effect of specifying relevant dose limits for different classes of person: adult employees aged 18 or over; trainees aged 16 to 18; and any other person. Thus the limit on effective dose is:

(a) 20 millisieverts a year for any person aged 18 or over who is an employee (as defined by regulation 2(2)(b) of these Regulations), except as provided by regulation 11(2) and Schedule 4, Part II (see paragraphs 200-205);

(b) 6 millisieverts a year for trainees aged 16 to 18 (see definition in regulation 2(1)); and

(c) 1 millisievert a year for any other person, eg members of the public and employees under 18 years of age who are not trainees (but see paragraphs 196-197 about special cases in which doses to members of the public may be averaged over five years).

Dose limits do not apply to people who are 'comforters and carers' or people undergoing a medical exposure (see regulation 3).

190   The limits only relate to exposures received by individuals in that class. The exposure of employees while they are at work is considered separately from any exposure they might receive outside working hours. Therefore the limits for adult employees will only apply to occupational exposures. Exposures received by an individual while not at work will be subject to the separate limits applying to any other person.

191   Employers are responsible for ensuring that dose limits are not exceeded for their employees, including trainees. Radiation employers should ensure that dose limits are not exceeded for people other than their own employees. However, this duty only relates to exposures resulting from their own work (see regulation 4(1)). Nevertheless, they will need to take into account doses already received by these people as a result of work with ionising radiation on other sites.

192   Exposures received as a result of natural background radiation at normal levels are not considered in determining compliance with dose limits. Exposures resulting from work with any radioactive substance containing naturally occurring radioactive materials do have to be considered (see regulation 3(1)(c) and ACOP advice in paragraph 11).

Assessing dose to the skin

193   Doses to organs and tissues are normally averaged over their volume for estimating effective dose and equivalent dose. The main exception is the skin, where the dose equivalent to the skin from external radiation, contamination on the skin or clothing and any incorporated radionuclides (in wounds) should be averaged over an area no greater than 1 cm^2 for comparison with the dose limit. For external radiation this will require careful consideration of the appropriate type and location of any dosemeter worn by an employee for assessing compliance with this limit. The aim should be to check that no area of the skin receives a dose exceeding the dose limit.

Additional dose limit for women of reproductive capacity

194   Any employer of a female employee who is (or may become) a classified person will need to consider if she will be exposed to ionising radiation at such a rate that she is likely to receive a dose to the abdomen exceeding 13 millisieverts in any consecutive three-month period. Where the working conditions of a female employee make it likely that she will receive a dose to the abdomen at this rate, the employer must record that information in her health record (see Schedule 7(d)). In such cases, the appointed doctor will need to consider whether she is in fact a woman of reproductive capacity and should be made subject to the additional dose limit for a woman of reproductive capacity (regulation 24 and Schedules 4 and 7(h)). This additional limit need only be considered where the appointed doctor has made a relevant entry in the health record of a female employee.

Disapplication of dose limits for comforters and carers

195   Comforters and carers are defined in regulation 2(1). As explained in paragraphs 127 and 128, they will normally be relatives or friends of the patient; they are not health-care employees. The exemption from dose limits can only apply where the individuals concerned knowingly and willingly accept the risk involved in providing comfort and support (see paragraph 129).

Dose limit for members of the public - exposures arising from patients undergoing medical diagnosis or treatment

196 Members of the public who may be significantly exposed to ionising radiation from being in close proximity to a patient receiving a medical exposure but who cannot be treated as comforters and carers are subject to the dose limits in Schedule 4 for 'any other person'. Such people may include friends who are unaware of the exposure they receive, and children. Children are unable to give their consent and other relatives or friends will not have the opportunity to do so unless they visit the hospital. In general, doses received by relatives and friends in this situation are likely to be small but difficulties could arise in particular situations without some flexibility in the application of the public dose limit of 1 millisievert a year effective dose.

197 In such cases, a hospital or clinic may take into account the provision in paragraph 7 of Schedule 4 for averaging the limit on effective dose for members of the public over a five-year period (5 millisieverts over five years). This provision should enable the hospital or clinic to make sensible arrangements for the release of patients who have been administered with a radiopharmaceutical and for restricting the exposure of relatives and friends. The British Institute of Radiology has published advice[25] on the timing of the release of such patients from hospital and the use of appropriate precautions. This advice relates to single administrations of radiopharmaceuticals and takes into account the restriction of consequential doses to members of the public.

Dose limit for members of the public - other exposures

198 The specific requirements for authorisation under the Radioactive Substances Act 1993,[6] which are enforced by the relevant environment agency, will generally ensure that doses to members of the public arising from authorised discharge and disposal of radioactive material satisfy the public dose limit. The responsible employer will need to ensure that any additional exposure from direct external radiation does not cause the dose limit to be exceeded for the relevant critical group.

Overexposures

199 The term 'overexposure' is defined in regulation 2(1). Regulation 25 applies when a radiation employer is informed or suspects that a person is likely to have received an overexposure. Regulation 26 provides an additional restriction for the continued exposure of any employee who has received an overexposure. This restriction must be treated as if it were a new dose limit for the remainder of the calendar year (or five-year period for those subject to a dose limit under regulation 11(2)).

Dose limitation for employees in special cases

(2) Where an employer is able to demonstrate in respect of any employee that the dose limit specified in paragraph 1 of Part I of Schedule 4 is impracticable having regard to the nature of the work undertaken by that employee, the employer may in respect of that employee apply the dose limits set out in paragraphs 9 to 11 of that Schedule and in such case the provisions of Part II of the Schedule shall have effect.

200 In some cases, because of the special nature of the work undertaken by an employee, it may not be practicable to comply with the annual limit of 20 millisieverts a year for adult employees. This situation may arise where there are skilled tasks which need to be undertaken by key specialist staff,

including foreign nationals. Where the employer can demonstrate that this is the case, the provisions in Part II of Schedule 4 will apply.

201 Any employer wishing to apply the special dose limit of 100 millisieverts in five years (and no more than 50 millisieverts in any single year) for an individual employee under regulation 11(2) and Schedule 4 Part II would have to establish a need. The employer would normally consider the work that individual is likely to do in the year ahead and make a judgement as to whether or not that employee's annual dose would exceed 20 millisieverts, based on predicted doses. The judgement would take account of past experience and any plans for restricting future exposures from particular jobs as far as is reasonably practicable. The employer would also need to be satisfied that the employee's dose would not exceed the five-year limit (or any other relevant dose limit) (see paragraph 12 Schedule 4).

202 The first five-year period for this limit starts on 1 January 2000. Doses received in any year before the year 2000 are not considered in the application of the limit.

203 The choice of this five-year dose limit for any particular employee is subject to a number of preconditions, which are set out in paragraph 13 of Schedule 4. These include: consultation with the RPA and with the affected employees; provision of information to the affected employees and the ADS; and giving prior notice to HSE. Employers may also need to consult any appointed safety representatives.

204 There are further conditions imposed once the five-year dose limit has been applied to an employee. There is a duty to investigate any suspected exposures exceeding 20 millisieverts in a calendar year and to notify HSE (paragraph 14 of Schedule 4). There is also a duty to review whether the five-year limit is still appropriate at least once every five years, and there are restrictions on reverting to an annual basis for the dose limit for that employee. The employer must record the reasons for choosing a five-year dose limit. The last three conditions are set out in paragraphs 15-17 of Schedule 4. Paragraphs 18 and 19 of Schedule 4 permit HSE to override the employer's decision, for example requiring the employer to apply an annual limit of 20 millisieverts to the individual(s) concerned. However, this action is subject to appeal to the Secretary of State.

205 The purpose of the investigation required by paragraph 14 of Schedule 4 is mainly to check that any future exposure arising from the work is unlikely to result in an overexposure, ie that the five-year dose limit will still be met. If the investigation shows that this is not the case, the employer should consider what action is necessary to change the working conditions of the individual. Note that if the individual has received an overexposure, an investigation under regulation 25 will be needed. There is no need for two investigations if an employee is made subject to the special five-year limit. The provisions of regulation 26 for any overexposure will apply to both the single-year and five-year dose limitation periods.

Regulation 12

Contingency plans

(1) Where an assessment made in accordance with regulation 7 shows that a radiation accident is reasonably foreseeable (having regard to the steps taken by the radiation employer under paragraph (3) of that regulation), the radiation employer shall prepare a contingency plan designed to secure, so far as is reasonably practicable, the restriction of exposure to ionising radiation and the health and safety of persons who may be affected by such accident.

51

Nature and scope of the plan

206 The purpose of the plan is to restrict any exposure that arises from an accident both to the employees themselves and to others, including emergency services personnel, who may be affected by it. The accident assessment and the reasonably practicable steps flowing from it, as required by regulation 7, provide the platform from which to build the plan.

207 Accident scenarios can range from the highly radiologically significant, perhaps involving industrial radiography or medical radiotherapy sources, to less serious events involving spillages of small quantities of radioactive materials. But the need for a contingency plan does not only depend on the accident's scale. Even in small-scale accidents, people are put under pressure and may well make incorrect judgements. The process of planning requires thought to be given beforehand to the correct courses of action to take in an accident situation, the recording of those actions in a plan and the training of those identified to implement the plan. Normally, the radiation protection adviser should be consulted about the plan (see ACOP advice in paragraph 217).

208 However, planning is only needed for reasonably foreseeable accidents. For an accident scenario to be just about credible is not sufficient. There must be reasonable grounds for believing it could occur.

209 The involvement of radioactive substances in fires is normally reasonably foreseeable and therefore subject to contingency planning. Fires can lead to dispersal of radioactive material which is otherwise well contained, or can melt lead shielding leading to substantial increases in dose rates.

210 In some scenarios considered in the plan (eg fires), the emergency services will become involved. Part of the planning process therefore involves consultation with these organisations, to ensure that they know and are content with the role they will be expected to play, and so that the radiation exposure of their own employees can be controlled.

211 Certain employers holding more than specified amounts of radioactive substances will also need to consider the requirements of regulations 26 and 27 IRR85 until they are replaced by separate regulations on emergency preparedness (see regulation 41(3)).

Content of the plan

212 The level of detail in the plan will need to reflect the circumstances anticipated. Some plans may be generic where the same operations are conducted in different places at different times but where the exposure conditions arising from a potential accident are likely to be similar. Site radiography and transport are two obvious examples.

213 In particular, the plan should normally identify:

(a) which postholders are responsible for putting the plan into effect;

(b) what immediate actions for assessing the seriousness of the situation will be necessary, for example the use of suitable radiation and contamination monitors;

(c) what immediate mitigating actions need to be taken, for instance in clearing the accident area and establishing temporary means of preventing access to that area;

(d) what personal protective equipment will be needed and where it may be found;

(e)   what personal dosimetry requirements there are for people involved in controlling the accident;

(f)   what training of personnel is required;

(g)   how to obtain radiation protection expertise so that proper judgements can be made about the seriousness of the situation and the measures necessary to recover from it;

(h)   under what circumstances to summon the emergency services; and

(i)   what dosimetry follow-up is needed so that all the people affected by the accident are identified and provision is made for their dose assessment.

(2)   The radiation employer shall ensure that -

(a)   where local rules are required for the purposes of regulation 17, a copy of the contingency plan made in pursuance of paragraph (1) is identified in those rules and incorporated into them by way of summary or reference;

(b)   any employee under his control who may be involved with or may be affected by arrangements in the plan has been given suitable and sufficient instructions and where appropriate issued with suitable dosemeters or other devices obtained in either case from the approved dosimetry service with which the radiation employer has entered into an arrangement under regulation 21; and

(c)   where appropriate, rehearsals of the arrangements in the plan are carried out at suitable intervals.

214   The choice of a suitable dosemeter or other device will need to take account of the types of radiation and the dose rates likely to be encountered. While a dosemeter provided for routine dose assessment will often be suitable, other types, for instance allowing immediate reading or giving high dose rate warning, may be more appropriate in some circumstances. Regulation 23 sets out requirements for dosimetry in cases of accidents and other incidents.

215   Considerations for judging the appropriateness of rehearsing the arrangements in the plan are:

(a)   the potential severity of the accident;

(b)   the likely doses that could be received by employees or others;

(c)   the complexity of the plan;

(d)   the number of people likely to be involved in its implementation; and

(e)   the involvement of the emergency services.

Regulation 13

Radiation protection adviser

(1)   Subject to paragraph (3), every radiation employer shall consult such suitable radiation protection advisers as are necessary for the purpose of advising the radiation employer as to the observance of these Regulations and shall, in any event, consult one or more suitable radiation protection advisers with regard to those matters which are set out in Schedule 5.

(2)   Where a radiation protection adviser is consulted pursuant to the requirements of paragraph (1) (other than in respect of the observance of that paragraph), the radiation employer shall appoint that radiation protection adviser in writing and shall include in that appointment the scope of the advice which the radiation protection adviser is required to give.

(3)   Nothing in paragraph (1) shall require a radiation employer to consult a radiation protection adviser where the only work with ionising radiation undertaken by that employer is work specified in Schedule 1.

216   To be suitable, a radiation protection adviser will need to possess the specific knowledge, experience and competence required for giving advice on the particular working conditions or circumstances for which the employer is making the appointment.

217   In addition to the specific matters set out in Schedule 5, radiation employers are required to consult a radiation protection adviser where advice is necessary for the observance of the Regulations. This should normally include:

(a)   the risk assessment required by regulation 7;

(b)   the designation of controlled and supervised areas as required by regulation 16, except where there is good reason to consider that such areas are not required, for example based on advice from the supplier of the radiation source or written guidance from an authoritative body;

(c)   the conduct of the various investigations required by the Regulations;

(d)   the drawing up of contingency plans required by regulation 12;

(e)   the dose assessment and recording required by regulation 21; and

(f)   the quality assurance programme in respect of medical equipment or apparatus required by regulation 32.

General advice on RPAs

218   The system of providing radiation protection advice under the Regulations places the onus on the individual (or body) wishing to act as a radiation protection adviser (RPA) to demonstrate that they meet HSE's criteria of competence. The radiation employer's responsibility, set out in regulation 13, is then to select from such RPAs one or more who have suitable knowledge and experience for the employer's type of work.

HSE criteria of competence

219 HSE has published a statement[8] on RPAs setting out criteria of competence for individuals and bodies intending to give advice as RPAs. This provides for recognition of core competence (basic RPA capability); employers will still need to select RPAs who have experience which is appropriate to the employer's business (see paragraph 222). Paragraph 4 gives further details about the HSE criteria of core competence for RPAs.

Transitional arrangements

220 For a transitional period up to the end of 2004, individuals or organisations that can show they have held appointment as an RPA under IRR85 can act as an RPA under IRR99, and therefore need not possess a certificate from an assessing body or a relevant N/SVQ.

Choosing a suitable RPA

221 The radiation employer must firstly ensure that any adviser consulted for statutory purposes conforms to the definition of an RPA, as described in the HSE statement on criteria of competence for RPAs.[8] For an individual, this means checking that the potential adviser either holds a valid certificate from an assessing body (in particular, checking that it is not time-expired), or has received a National/Scottish Vocational Qualification at Level 4 in Radiation Protection Practice within the previous five years. RPA bodies should be able to provide evidence of formal recognition by HSE (subject to paragraph 220).

222 Equally important, the RPA should be suitable in terms of possessing the requisite knowledge and experience relevant to the employer's type of work. The judgement about suitability would principally be derived from the appropriateness of the RPA's working history.

223 Employers should be aware that HSE acceptance of organisations as assessing bodies is limited to the purpose of establishing core competence of RPAs. However, assessing bodies are permitted to issue RPAs with certificates that carry information additional to that required by HSE, acknowledging specific areas of expertise that the individual possesses. When HSE considers the suitability of potential assessing bodies, no account will be taken of any arrangements to provide this additional acknowledgement of specific areas of expertise. Nevertheless, employers may find this additional information helpful in judging the suitability of an RPA.

224 On the other hand, in many instances the RPA cannot be expected to possess all the specialist knowledge required by the employer. So another important characteristic is their ability to recognise the limitations of their knowledge and experience in certain areas and have access to that specific information from other sources. For instance, while having a working understanding of the types of instrument to use in particular situations, they may well not be experts in instrumentation. Similarly, an RPA dealing with radon problems may well need access to expertise in ventilation to advise on adequate exposure restriction.

225 Organisations that employ a number of RPAs may decide to take on an individual who is currently an RPA and possesses the right attributes but has little experience of the employer's type of work. The employer may consider that such a person has the potential to be developed as a suitable RPA but is not suitable at present. The individual would then only be appointed as an RPA by the new employer at a later date, once that person had acquired the relevant experience.

Consulting an RPA

226   Where the radiation employer needs advice on compliance with the Regulations, then this must be sought through consultation with an RPA as defined in regulation 2(1). Such consultation is obligatory for the matters listed in Schedule 5, and also if the dose limitation scheme in Part II of Schedule 4 is to be invoked. Item 1 of Schedule 5 *Implementation of requirements as to controlled and supervised areas* should be read as covering all of the requirements in regulations 17-19. However, see paragraph 231 for exemptions from the requirement in regulation 13(1).

227   In addition, consultation with an RPA is normally expected for the matters listed in ACOP paragraph 217. In this respect, the RPA's involvement in the prior risk assessment required by regulation 7 (or reviewing an assessment made under regulation 3 of MHSWR[7]) could be especially valuable by identifying at the outset where the risks arise and what precautions and procedures are going to be necessary. Beyond this, there is a range of matters where, depending on the experience of the radiation employer, the expertise of the RPA may well be necessary. The matters include the selection and use of personal protective equipment, arrangements for outside workers, and training (Table 2 summarises the circumstances in which an RPA may need to be consulted). It would be prudent for the radiation employer to seek written confirmation from the RPA of any key advice as this would provide evidence of consultation.

Appointment of RPAs

228   Where a radiation employer intends to consult a suitable RPA for the purposes of these Regulations then the employer must also appoint that RPA in writing; appointments made under earlier regulations are not sufficient and a reappointment will be needed. However, an appointment is not required for an initial consultation, ie where the RPA is only asked to advise the employer as to whether, given the nature of the work with ionising radiation, formal consultation is necessary on the matters mentioned in regulation 13(1).

229   RPAs can be part-time or full-time employees or consultants, depending on the extent of demand for continuing advice. For smaller organisations, for example those whose work with ionising radiation is confined to using a gauging device containing a sealed source of ionising radiation, a consultant may well suffice. In this case, detailed advice required at the early stages of setting up might normally be followed by an access arrangement on an as-and-when basis. Conversely, a large or complex organisation may need a number of RPAs (or an RPA body) if the range of tasks, or scope and complexity of the work with ionising radiation is such that one person could not reasonably provide the breadth of advice or the functions required. These RPAs might possibly have different individual areas of expertise for different purposes. Either way, written appointment is intended to formalise the terms of the arrangement between the RPA and the employer, in particular by specifying the scope of the advice the RPA is required to give. A letter would often be adequate for this purpose and the appointment remains until subsequently terminated.

Availability of RPA advice

230   RPAs appointed on a continuing basis should usually be available for consultation whenever required under the access arrangement agreed with the employer, though they do not need to be present every time that work with ionising radiation takes place.

Table 2 Consulting a suitable RPA on specific matters

Required by regulations 13(1) & 31(2) and Schedules 4 & 5	Normally expected to conform to ACOP advice
Prior examination of plans for installations and acceptance of physical control measures for sources being introduced into service (regulation 8)	Risk assessment (regulation 7)
Critical examinations by installers or erectors of articles (regulation 31)	Designation of controlled and supervised areas as appropriate (regulation 16)
The periodic examination and testing of physical control measures and checking of systems of work (regulations 8 and 10)	Conduct of investigations (regulations 8(7), 22, 25, 30, 32 and Schedule 4)
Implementing requirements for controlled and supervised areas such as demarcation, signs, adequate monitoring, written arrangements for entry by non-classified persons and local rules (regulations 17, 18 and 19)	Drawing up contingency plans (regulation 12)
Requirements for regular calibration of instruments provided for monitoring levels of ionising radiation (notably in designated areas) and regular checking that such equipment is serviceable and correctly used (regulation 19)	Dose assessment and dose recording (regulation 21)
Use of the special system of dose limitation for employees (regulation 11(2) and Schedule 4 Part II)	QA programme for equipment used in connection with medical exposures (regulation 32)
Other situations where, in the particular circumstances, RPA advice is necessary for the observance of the Regulations, eg selection and use of adequate and suitable PPE (regulation 8); training (regulation 14); and arrangements for outside workers (regulations 18 and 21)	

Exemption from the need to appoint an RPA

231 Employers whose work with ionising radiation comes within any of the descriptions in Schedule 1 are not required to consult and do not need to appoint an RPA. However, they may nevertheless wish to consult an RPA, at least initially, for checking or reassurance purposes.

Information and facilities for the RPA

(4) The radiation employer shall provide any radiation protection adviser appointed by him with adequate information and facilities for the performance of his functions.

232 Radiation employers who need advice in relation to plans for off-site emergencies should provide, or may arrange to share, a specialised radiation protection unit. Such units should be distinct from production and operational units and authorised to perform radiation protection tasks.

233 Employers should make sure that their RPAs have access to all the information and facilities that they need to perform effectively. The information should include a clear statement of the scope of the advice each RPA is required to give. The facilities may need to include appropriate support services (eg secretarial), unless the RPAs provide their own. Where there is a potential for emergencies with off-site consequences, the ACOP advises that specialised radiation protection units should normally be provided to support the RPA(s).

Regulation 14

Information, instruction and training

Every employer shall ensure that -

(a) those of his employees who are engaged in work with ionising radiation are given appropriate training in the field of radiation protection and receive such information and instruction as is suitable and sufficient for them to know -

(i) the risks to health created by exposure to ionising radiation;

(ii) the precautions which should be taken; and

(iii) the importance of complying with the medical, technical and administrative requirements of these Regulations;

(b) adequate information is given to other persons who are directly concerned with the work with ionising radiation carried on by the employer to ensure their health and safety so far as is reasonably practicable;

Who needs information, instruction and training?

234 All employees involved in the work with ionising radiation, including management, will need training to help develop and sustain a commitment to restricting exposure wherever this is reasonably practicable. The employer will usually need to provide training to ensure employees are competent where a system of work or personal protective equipment is provided to restrict exposure (as required by regulation 8). Training will also be needed where the employer arranges for employees to perform particular functions required by these Regulations, for example:

(a) to act as a radiation protection supervisor (regulation 17);

(b) to make entries in radiation passbooks for outside workers (regulation 18); or

(c) to monitor radiation levels for controlled or supervised areas (regulation 19).

Some employees may not be closely involved with the work but need suitable information or instruction to avoid being unnecessarily exposed to ionising radiation. The duty under this regulation complements the general duties on information, capabilities and training in regulations 10 and 13 of MHSWR.[7]

235 Employees involved in work with ionising radiation need to be made aware of the main risks, including the risk of accidental exposures, and the control measures provided to prevent or reduce those risks. Where appropriate, key staff should have adequate information about how and when to consult the RPA or other experts such as approved dosimetry services or the appointed doctor.

236 The duty in regulation 14 is placed on all employers in relation to their employees; it is not limited to employers who are undertaking the work with ionising radiation (radiation employers). If a contractor carries out work other than work with ionising radiation on the site of a radiation employer, both employers will be expected to share information, under regulation 15. Effective co-operation should ensure that the contractor can inform their own employees about any radiation risks associated with their work on the site and any preventative measures they must take to avoid those risks.

What is required?

237 The need for information, instruction and training applies to all people who may be affected. The different categories of person who require information, etc include:

(a) classified persons (see regulation 20);

(b) outside workers (see regulation 18(2)(b));

(c) people who enter controlled areas under written arrangements (see regulation 18(3));

(d) other workers in the establishment who may, for example, need to recognise warning signs;

(e) safety representatives;

(f) radiation protection supervisors (RPS) (see regulation 17), who need to know enough about radiation protection principles and procedures, the requirements of the Regulations and the arrangements in local rules to enable them to supervise the work safely;

(g) employees who undertake monitoring for the purposes of regulation 19, for example to use the right monitoring instrument, check that it is functioning correctly and make adequate measurements;

(h) supervisors and managers with specific responsibilities under these Regulations.

238 Training should be appropriate to the nature of the work and the needs of the individual. Individuals should be aware of when they need to seek help and where they should find it. Employers are advised to consult their RPAs when planning who needs what information, instruction and training. Records of training may be helpful to enable the employer to identify who needs further training/refresher training.

Information for female employees who may become pregnant or start breast-feeding

Every employer shall ensure that -

> *(c) those female employees of that employer who are engaged in work with ionising radiation are informed of the possible risk arising from ionising radiation to the foetus and to a nursing infant and of the importance of those employees informing the employer in writing as soon as possible -*
>
> > *(i) after becoming aware of their pregnancy; or*
> >
> > *(ii) if they are breast feeding.*

239 Female employees should be urged to notify their employer in writing as soon as they become pregnant or if they are breast-feeding, so that the employer can put in place any special protection required under regulation 8(5) (unless working conditions are such that no special protection would be necessary).

240 Information on risks from exposure to ionising radiation, particularly in respect of the possible risk to the foetus, could include any relevant guidance from HSE or the appointed doctor.

Co-operation between employers

Where work with ionising radiation undertaken by one employer is likely to give rise to the exposure to ionising radiation of the employee of another employer, the employers concerned shall co-operate by the exchange of information or otherwise to the extent necessary to ensure that each such employer is enabled to comply with the requirements of these Regulations in so far as his ability to comply depends upon such co-operation.

Employers sharing the same workplace

241 If two employers share the same workplace, whether on a temporary or permanent basis, those employers will have duties under regulation 11 of MHSWR[7] to co-operate with each other. The aim of co-operation should be to co-ordinate the measures they take to comply with statutory duties and to inform each other of the risks to employees arising out of their work (note also the need for site employers to pass on comprehensible information about risks under regulation 12 of MHSWR). Normally compliance with that regulation should be sufficient to satisfy the requirements of regulation 15 of these Regulations. Information that might need to be shared would include details relating to controlled areas that could be entered by employees of either employer and relevant contingency arrangements for action to prevent or mitigate the consequences of any radiation accident (see regulations 7(3), 12 and 18(2)).

242 As regards work in buildings with radon atmospheres subject to these Regulations, it is important for the employer controlling the premises to notify other employers working in the building.

Sharing dose information

243 Where an employee has several employers, those employers may need to co-operate to ensure that dose limits are applied to the total dose received by the worker. For example, in the health sector there are several groups of personnel (such as cardiologists) who may have multiple employers (eg an

NHS trust and one or more private hospitals) or who may act in a self-employed capacity for some of the year. It is important for employers to share information on the total dose received otherwise they will have difficulties in making the correct decisions on the need for classification and ensuring compliance with relevant dose limits.

Outside workers

244 Employers of outside workers should normally find out from the employer in control of the work area what risks and additional training needs are associated with the services the outside worker will perform, well before those services are due to start. Generally, it is sensible to share this information at the stage when contracts and arrangements are still being discussed. This should help ensure that the right workers are chosen to carry out the services and that, before they start, those workers are given the relevant information and training to enable them to protect themselves and others properly. Employers in control of the work area will also need information about the outside worker (eg dose history and training needs) before they start work on site (see regulation 18(2)(b)). In most cases, employers will find it helpful to consult their RPAs on this matter.

245 Employers of outside workers will usually require the following basic information:

(a) the detail of the actual work to be done;

(b) the type of likely radiation exposure;

(c) an estimate of the total dose likely to arise from the work;

(d) the work procedures that will be required to keep doses as low as reasonably practicable (including any use of special protective equipment);

(e) any local restrictions that will be applied;

(f) the local rules that apply in the other employer's site (including emergency arrangements and contingency plans);

(g) radiation protection supervisor (RPS) appointments;

(h) any relevant dose constraints, special dose limits or declaration of pregnancy or breast-feeding.

The amount of detail will depend on the complexity and duration of the work. Employers of outside workers will also need to inform the employer in control of the work area about any risks arising from work with ionising radiation they intend to undertake.

246 Employers will need to know whether additional dose assessments will be necessary for their outside workers. For example, assessment of external radiation exposures to the eye or an extremity, or of internal exposure, may be necessary. If additional dose assessment is needed, the employers should agree between them who will make the necessary arrangements. The outside worker's employer will also need to be aware of the methods of dose estimation that the other employer will use. In the event of an incident, additional dose estimates may be required (see regulation 23).

247 If the work will take place outside normal hours, the employers may need to exchange information on emergency contacts.

PART IV DESIGNATED AREAS

Designation of controlled or supervised areas

Regulation

16(1)

(1)   Every employer shall designate as a controlled area any area under his control which has been identified by an assessment made by him (whether pursuant to regulation 7 or otherwise) as an area in which -

(a)   it is necessary for any person who enters or works in the area to follow special procedures designed to restrict significant exposure to ionising radiation in that area or prevent or limit the probability and magnitude of radiation accidents or their effects; or

(b)   any person working in the area is likely to receive an effective dose greater that 6mSv a year or an equivalent dose greater than three-tenths of any relevant dose limit referred to in Schedule 4 in respect of an employee aged 18 years or above.

ACOP

248   Special procedures should always be necessary to restrict the possibility of significant exposure, and therefore employers should designate controlled areas, in cases where:

(a)   the external dose rate in the area exceeds 7.5 microsieverts per hour when averaged over the working day;

(b)   the hands of an employee can enter an area and the 8-hour time average dose rate in that area exceeds 75 microsieverts per hour;

(c)   there is a significant risk of spreading radioactive contamination outside the working area;

(d)   it is necessary to prevent, or closely supervise, access to the area by employees who are unconnected with the work with ionising radiation while that work is under way; or

(e)   employees are liable to work in the area for a period sufficient to receive an effective dose in excess of 6 millisieverts a year.

249   In addition, an area should be designated as a controlled area if the dose rate (averaged over a minute) exceeds 7.5 microsieverts per hour and:

(a)   the work being undertaken is site radiography; or

(b)   employees untrained in radiation protection are likely to enter that area, unless the only work with ionising radiation involves a radioactive substance dispersed in a human body and none of the conditions in the previous paragraph apply.

In this context, site radiography means any radiography of inanimate objects other than that which is carried out in an enclosure or cabinet that restricts the dose rate (averaged over a minute) outside the enclosure to 7.5 microsieverts per hour.

16(1)

Responsibility for designating an area

250   The responsibility for designating such areas rests with the employer in control of that area. Usually this will be the employer who is in overall control of the site, eg the main radiation employer. Where that employer assigns temporary control of the area to a contractor, that contractor has the responsibility for deciding whether or not to designate the area as a controlled area. Contractors will need to co-operate with site employers, either to inform them about the extent of any controlled area they create, or to pass on information about the risks arising from their work with ionising radiation.

251   Paragraph 217 advises that the employer should normally consult an RPA about the need to designate a controlled or supervised area.

Purpose of designating controlled areas

252   The main purpose of designating controlled areas is to help ensure that the measures provided under regulations 7(3) and 8(1) are effective in preventing or restricting routine and potential exposures. This is achieved by controlling who can enter or work in such areas and under what conditions. Normally, controlled areas will be designated because the employer has recognised the need for people entering the area to follow special procedures. Such procedures could take the form of a detailed system of work which sets out how the tasks should be undertaken in a way that restricts significant exposure.

Designation as a controlled area on the basis of special procedures

253   The risk assessment undertaken by the employer will indicate where special procedures are necessary to restrict exposure, in addition to the physical control measures required by regulation 8. If these procedures are specific to an area and require particular instructions to be followed by those who enter or work in the area (or untrained people to be excluded unless under close supervision), the area will probably need to be designated. The ACOP guidance in paragraphs 248 and 249 advises on situations where a controlled area should normally be designated.

254   Where the employer considers it unnecessary to have special procedural controls, having undertaken a risk assessment, the area must nevertheless be designated as a controlled area if any person is likely to receive an annual effective dose in excess of 6 millisieverts, or an annual equivalent dose greater than one of the other values specified in regulation 16(1)(b), as a result of work in that area.

255   The employer's assessment may show that particular procedures are necessary to prevent accidental exposures when people enter high dose rate shielded enclosures or plant, eg it is necessary for employees to follow a defined procedure involving the use of a suitable dose rate meter to check that a radiation source is safe before entry. The assessment may also indicate other situations where anyone entering or working in an area should follow special procedures, including work areas where the dose rate is less than that indicated in paragraphs 248 and 249. If special procedures are necessary, the area should be designated as a controlled area notwithstanding the fact that the dose rate is below 7.5 microsieverts per hour.

256 In deciding whether or not a controlled area is needed, employers will want to take account of such factors as:

(a) which people are likely to need access to the area;

(b) the level of supervision required;

(c) the nature of the radiation sources in use and the extent of the work in the area;

(d) the likely external dose rates to which anyone can be exposed;

(e) the likely periods of exposure to external radiation;

(f) the physical control methods already in place, such as permanent shielding and ventilated enclosures;

(g) the importance of following a procedure closely in order to avoid receiving significant exposure;

(h) the likelihood of contamination arising and being spread unless strict procedures are closely followed;

(i) the need to wear personal protective equipment in that area; and

(j) maximum doses estimated for work in the area.

257 In addition to the circumstances described in paragraphs 248 and 249, the employer may find it necessary to designate an area as controlled if:

(a) access is foreseeable to that area by people, such as office staff, whose work does not normally involve ionising radiation (see ACOP advice in paragraph 60);

(b) normal control measures for an area have to be suspended for work such as maintenance or source changing;

(c) people are likely to be exposed to significant levels of surface or airborne contamination in the area, in excess of appropriate derived working levels or derived air concentrations;

(d) respiratory protective equipment must be worn while working in the area.

Circumstances where designation is unlikely to be needed

258 It is not necessary for the employer in control to designate an area as a controlled area simply because people work under a general system that reflects good practice in that sector. This kind of system of work is not regarded as a 'special procedure'. For example, a controlled area would not normally be required where:

(a) work is routine and special precautions are not required, for example work in the vicinity of a fixed radiation gauge (except maintenance work); or

(b) work is carried out with low levels of radionuclides of low radiotoxicity inside efficiently ventilated enclosures (eg fume cupboards) or on a laboratory bench and only routine precautions are expected, such as the use of lined trays to contain spillage and the use of disposable protective gloves.

259   Places which cannot physically be entered do not need to be designated. It is not necessary to designate an area as a controlled area if it is not reasonably foreseeable that a person, or part of a person, will enter or be present in that area.

260   Designation of an area is not required if the only person in that area who is exposed to ionising radiation will be a person undergoing medical examination or treatment (see regulation 3(3)). However, the employer needs to consider the possibility of exposures, including accidental or unintended exposures, of other members of the public and members of staff (see also paragraphs 550-553 concerning potential exposures resulting from defects or malfunctions in equipment used for medical exposures).

Designation in the case of radionuclides in the human body

261   Designation will probably be necessary in a few limited cases, for example where a patient remains in a hospital or clinic after the therapeutic administration of a radiopharmaceutical and:

(a)   the work with ionising radiation involves a radioactive substance dispersed in a human body where that substance emits gamma rays and the product of activity and total gamma energy per disintegration exceeds 150 MBq. MeV; or

(b)   the patient is undergoing brachytherapy using a gamma source or a high energy beta ray source.

Designation of an area taking account of physical features

262   In determining the extent of any controlled area, the employer may need to take account of the physical boundaries, such as walls and partitions around the working area. If it is more convenient to use these boundaries (eg because of the need to control access), then they may be used rather than a smaller part of the area where dose rates or contamination levels are significant. However, these boundaries should not be too remote from the area of concern to enable proper control to be exercised. Once such an area has been designated, it is subject to all the legal requirements applying to controlled areas under regulations 17, 18 and 19. However, it is not necessary to designate the whole of a room as a controlled area provided that the necessary restrictions can be applied to that part of a room or laboratory where it is necessary to prevent or restrict access.

Temporary de-designation of controlled areas

263   If the periods during which work with ionising radiation takes place are clearly defined, follow a regular pattern, or are only intermittent, the employer may wish to de-designate on a regular basis, for example to allow cleaners to have routine access where this would be appropriate. This may be done provided sufficient steps are taken to remove the need for designation of the area, for example any X-ray generator is isolated from the power supply or any radioactive substances are removed or otherwise made safe. These steps will need to be summarised in local rules prepared in compliance with regulation 17(1) (see paragraphs 272-285).

Temporary designation of an area as controlled for a particular task, for example source changing or maintenance

264   An employer may decide it is unnecessary to designate an area as a controlled area because employees do not enter that area and physical

65

safeguards prevent accidental exposures. However, if contractors have to enter the area for particular tasks, such as maintenance, it may be necessary to designate such areas temporarily under specified conditions (see paragraph 250).

Controlled areas around packages during transport or movement on site

265   In some cases, the person accompanying a package containing a radioactive substance may be in an area which merits designation as a controlled area. However, bystanders are not likely to be in that situation. Under normal conditions of transport or movement around a site, it is unlikely that a controlled area would exist outside the edges of a package if the radioactive substance is packaged in accordance with the requirements of the International Atomic Energy Agency in relation to the safe transport of radioactive materials (see also paragraphs 508-511).

Designation on the basis of annual dose

266   In practice, it is often difficult to predict annual doses received by employees from knowledge of dose rates in working areas, because:

(a)   dose rates are seldom constant over long time periods and within the physical boundaries of areas;

(b)   there are significant variations in the pattern of work for individuals; and

(c)   the duration of an individual's exposure in the areas may be difficult to estimate.

Consequently, the expected annual dose is not likely to be the main criterion in most cases for deciding whether an area needs to be designated as a controlled area. One exception might be areas in radon-affected workplaces where high radon levels are known to occur and no special procedures need to be followed by employees. Also, where employees work for about 2000 hours a year in an area where the external dose rate routinely exceeds 3 microsieverts per hour, that area may need to be designated as a controlled area because that individual would be likely to receive a dose greater than 6 millisieverts a year.

(2)   An employer shall not intentionally create in any area conditions which would require that area to be designated as a controlled area unless that area is for the time being under the control of that employer.

Intention to create conditions

267   The area concerned may be under the control of another employer or outside the boundary of a site. In either case, the purpose of this provision is that radiation employers do not plan or undertake work that gives rise to external dose rates or levels of contamination which require access to be restricted in an area they do not control. A typical example would be site radiography work adjacent to a footpath where people walking past the area could be significantly exposed to ionising radiation. However, if these conditions arise in an area as a result of an unforeseen accident, the employer would not be considered to have intentionally created such conditions.

Designation of an area after an accident

268 An unforeseen accident might create conditions which would warrant the designation of a controlled area, but for the operation of regulation 16(2), in a place where the employer does not normally have control. In such cases, as part of a contingency plan under regulation 12, that employer should normally try to have the access to that area restricted until the situation returns to normal or until the emergency services take over control of it.

Designation of supervised areas

(3) An employer shall designate as a supervised area any area under his control, not being an area designated as a controlled area -

(a) where it is necessary to keep the conditions of the area under review to determine whether the area should be designated as a controlled area; or

(b) in which any person is likely to receive an effective dose greater than 1mSv a year or an equivalent dose greater than one-tenth of any relevant dose limit referred to in Schedule 4 in respect of an employee aged 18 years or above.

269 The decision to designate an area as a supervised area depends both on the assessment of likely doses in that area and the probability that conditions might change. For example, an area may need to be kept under review and therefore designated as a supervised area because of the possibility that radioactive contamination might spread. However, it will not be necessary to designate a supervised area outside every controlled area. For example, if a controlled area has been designated on the basis of external dose rate, and conditions in adjacent areas are unlikely to alter significantly, a supervised area will not be necessary unless a person is likely to receive a dose in excess of 1 millisievert a year in those adjacent areas.

270 In laboratories where only small quantities of unsealed radioactive substances are used, it may not be appropriate to designate the whole room as a controlled area to ensure that specific procedures are followed by those who enter or work there. In such a laboratory, however, there will be general arrangements for preventing spillages as far as possible and for cleaning up any contamination arising from a foreseeable spillage. Normally, the employer should designate at least part of the laboratory as a supervised area if contamination could build up over a period of some weeks as a result of not following these arrangements.

271 Employers can choose boundaries for the supervised area which are convenient. Once such an area has been designated it is subject to all the legal requirements applying to supervised areas.

Local rules and radiation protection supervisors

(1) For the purposes of enabling work with ionising radiation to be carried on in accordance with the requirements of these Regulations, every radiation employer shall, in respect of any controlled area or, where appropriate having regard to the nature of the work carried out there, any supervised area, make and set down in writing such local rules as are appropriate to the radiation risk and the nature of the operations undertaken in that area.

272 Written local rules should identify the key working instructions intended to restrict any exposure in that controlled or supervised area. The details given in these rules should be appropriate to the nature and degree of the risk of exposure to ionising radiations. The rules should cover work in normal circumstances and also the particular steps to be taken to control exposure in the event of a radiation accident, as set out in the contingency plan required by regulation 12. Local rules for a controlled area should include a summary of the arrangements for restricting access into that area, including the written arrangements covering those who are not classified persons.

Responsibility for local rules

273 The responsibility for ensuring that local rules are prepared rests with the employer undertaking the work with ionising radiation (the radiation employer). This is because that employer has the overall responsibility for providing control measures to restrict exposure under regulation 8. Where more than one radiation employer works in a controlled area, each such employer will have a duty to prepare local rules, even if these are adapted from the site occupier's local rules.

Local rules for supervised areas

274 Local rules are likely to be appropriate for supervised areas where the radiation employer needs to instruct employees about general arrangements to prevent accidents or to restrict exposure in that area. Examples include: maintenance and cleaning of an area where unsealed sources are used; and arrangements for putting a contingency plan into effect in the event of an accident.

Nature of local rules

275 The main purpose of local rules is to set out the key arrangements for restricting exposure in a particular area. They can vary considerably in detail and format, depending on the complexity of the work with ionising radiation. Local rules can take the form of, or include, instructions, booklets or circulars. Although local rules are only required for controlled areas (and, where appropriate, supervised areas), employers may choose to have local rules that relate to work with ionising radiation on the whole site, not just in controlled and supervised areas. Indeed, radiation employers may wish to continue with existing rules, adapted as necessary, which were prepared to meet earlier legal requirements. In this case, the employer should ensure that the local rules used contain the required details, include relevant descriptions of the area(s) concerned, and are brought to the attention of the relevant groups of employees as required by regulation 17(3).

276 The radiation employer's general management arrangements for complying with the Regulations form part of the general health and safety arrangements required by regulation 5 of the MHSWR.[7] Those arrangements may set out management responsibilities for radiation protection and include arrangements for monitoring or auditing the measures established to comply with these Regulations. There is no need to duplicate those arrangements in local rules. However, the radiation employer may wish to cross-refer to the main features relevant to a particular area, or even to repeat or attach a copy of those general arrangements to a centrally held copy of the local rules (but see paragraph 279). Local rules might be held in an electronic system. However, readily available paper copies may be needed for the radiation protection supervisor (RPS) and staff working in the area, especially where the work is undertaken away from the base location of the employer.

277 The radiation employer must seek advice from the RPA about the preparation of local rules (one of the requirements for controlled and supervised areas covered by Schedule 5) and may also need to consult any established safety committee or appointed safety representative(s).

Essential contents of local rules

278 Local rules should contain at least the following information:

(a) the dose investigation level specified for the purposes of regulation 8(7);

(b) identification or summary of any contingency arrangements indicating the reasonably foreseeable accidents to which they relate (regulation 12(2));

(c) name(s) of the appointed radiation protection supervisor(s) (regulation 17(4));

(d) the identification and description of the area covered, with details of its designation (regulation 18(1)); and

(e) an appropriate summary of the working instructions, including the written arrangements relating to non-classified persons entering or working in controlled areas (regulation 18(2) and the ACOP advice in paragraph 272).

279 Where a radiation employer has detailed written working instructions contained within operations manuals or work protocols, it will usually be sufficient for the local rules to refer to the relevant sections of these documents. However, the employer must ensure that the way in which these are summarised in local rules is adequate for the purposes of regulation 17.

Optional contents of local rules

280 The radiation employer may also find it useful to include a brief summary of, or refer to, the general arrangements in that area for:

(a) management and supervision of the work (see paragraph 276);

(b) testing and maintenance of engineering controls and design features, safety features and warning devices;

(c) radiation and contamination monitoring;

(d) examination and testing of radiation monitoring equipment;

(e) personal dosimetry; and

(f) arrangements for pregnant and breast-feeding staff.

281 The radiation employer may also wish to include information on relevant aspects of the health and safety management system, for example:

(a) details of significant findings of the risk assessment or where it can be found;

(b) a programme for reviewing whether doses are being kept as low as reasonably practicable and local rules remain effective;

(c) procedures for initiating investigations etc;

(d) procedures for ensuring staff have received sufficient information, instruction and training; and

(e) procedures for contact and consultation with the appointed RPA;

where these matters are not already covered in general health and safety arrangements required by regulation 5 of the MHSWR.[7] An annex to the local rules could be used for this information, to avoid obscuring the key features mentioned in paragraph 278.

Level of detail required

282 Local rules are intended to focus on arrangements that are needed in a particular area. The level of detail required in the written local rules will depend on the nature of the work in that area. Therefore, radiation employers will probably need to consider such matters as:

(a) the risk arising from the exposure in normal operations;

(b) the complexity of the work being done;

(c) the likelihood of a failure or accident and the magnitude of any resulting exposures; and

(d) the level of information, instruction and training given to those working in the area and the extent of their knowledge and experience.

These matters would normally be considered as part of the employer's risk assessment.

283 If, as a result of an accident, a particular operation could give rise to exposures approaching the investigation level provided under regulation 8(7), the radiation employer may need to specify all the critical parts of that operation in local rules and explain where the full instructions are held. Examples of these critical parts might include: monitoring of the source position at key stages; and hand-over procedures between shifts or between maintenance personnel and users. However, if the work is fairly routine (eg cleaning up minor spillages of radioactive materials on a laboratory bench), local rules may simply refer to detailed instructions held elsewhere in the area.

Making local rules effective

284 Local rules are likely to be effective if they:

(a) are brief and concentrate on the activities which give the greatest risks;

(b) focus on the working instructions to be followed by everyone to keep radiation doses as low as reasonably practicable;

(c) are 'local' by referring directly to work at a particular location and by taking account of the environment in and around the working area;

(d) contain clear instructions which reflect actual work practice rather than an ideal which is not attainable; and

(e)   are reviewed periodically to check that they remain relevant.

285   The radiation employer will need to strike a balance between the provision of necessary details and the ability of those affected to read and act on the local rules. If the main body of the local rules is too long, or unclear, people entering or working in the area may not read them; in which case they are not likely to be effective.

Regulation

17(2)

(2)   The radiation employer shall take all reasonable steps to ensure that any local rules made pursuant to paragraph (1) and which are relevant to the work being carried out are observed.

286   Regulation 5 of MHSWR[7] requires employers to put a management and supervisory system in place to help ensure the effectiveness of the health and safety measures. These measures will include the control measures necessary for compliance with the Ionising Radiations Regulations 1999,[1] supplemented by other steps to ensure that local rules are followed.

287   The radiation employer should consider what additional arrangements for supervision are necessary to comply with the general duty under regulation 5 of MHSWR.[7] For example, where a small team of employees spends long periods working away from the base location, it may be appropriate for a regional manager to arrange for periodic audits of the way the work is being carried out.

Regulation

17(3)

(3)   The radiation employer shall ensure that such of those rules made pursuant to paragraph (1) as are relevant are brought to the attention of those employees and other persons who may be affected by them.

288   Local rules should be available to employees and other people directly involved or likely to be affected at or near the area concerned, for example by displaying sections of the rules relevant to particular operations in, or immediately adjacent to, the area. They may be seen as a collection of parts relating to different areas so employees need only see relevant parts relating to the areas in which they work. Employees will then find it easier to take account of the particular working instructions appropriate to their work. A complete set of local rules for a site could be held for reference at one central point. This may be helpful to managers responsible for reviewing and updating the local rules (and for appointed safety representatives).

Appointing suitable radiation protection supervisors

Regulation

(4)   The radiation employer shall –

(a)   appoint one or more suitable radiation protection supervisors for the purpose of securing compliance with these Regulations in respect of work carried out in any area made subject to local rules pursuant to paragraph (1); and

17(4)

(b)   set down in the local rules the names of such individuals so appointed.

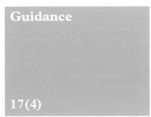

General advice on radiation protection supervisors

289   The appointment of a radiation protection supervisor (RPS) under this regulation supplements, but is not a substitute for, the general requirements in regulations 5 and 7 of MHSWR for monitoring and health and safety assistance. The RPS has a crucial role to play in helping to ensure adherence to the arrangements made by the radiation employer, and in particular

supervising the arrangements set out in local rules. However, the legal responsibility remains with the employer and cannot be delegated to the RPS.

Suitability for appointment as an RPS

290   A person's suitability for appointment as an RPS will depend both on a knowledge and understanding of the Regulations and local rules and an ability to exercise a supervisory role. Employees who are appointed as RPSs are unlikely to fulfil the role envisaged by this regulation unless they:

(a)   know and understand the requirements of the Regulations and local rules relevant to the work with ionising radiation;

(b)   command sufficient authority from the people doing the work to allow them to supervise the radiation protection aspects of that work;

(c)   understand the necessary precautions to be taken and the extent to which these precautions will restrict exposures; and

(d)   know what to do in an emergency.

291   RPSs should receive appropriate training under regulation 14 to fulfil the task adequately (see paragraph 234). In some cases, as little as a few hours of targeted training, for example from the RPA, may be sufficient. But the training will need to reflect the complexity of the work undertaken.

292   In general, RPSs will be employees of the employer undertaking the work with ionising radiation. They will usually be in line management positions, closely involved with the work being done, to allow them to exercise sufficient supervisory authority. In some situations, for example where a contractor is undertaking work on the site of another employer, it may be appropriate to appoint one of that site operator's employees as the RPS. Such an appointment might be appropriate where the site operator is a radiation employer and the contractor rarely undertakes work with ionising radiation. However, as the RPS undertakes a supervisory role, both employers may need to make suitable contractual arrangements for such an appointment.

Number of RPSs required

293   The role of the RPS is to exercise close supervision of the work on behalf of the employer, but it may not always be necessary for an RPS to be present all the time. The radiation employer will need to take account of the range and complexity of the work and the number of different locations to be covered in deciding how many RPSs are required. Cover should be sufficient for the work being done. In determining the number of RPSs required, the radiation employer will need to take account of factors such as peripatetic work away from the base location of the company, shift work and absence due to sickness, training and holidays.

Appointment of the RPS

294   Any RPS appointed by the radiation employer should be clear about the role they are expected to fulfil. The name of each RPS appointed has to be given in the local rules covering the areas they will supervise (see paragraph 278). The radiation employer may also wish to display the name of the relevant RPS on notice boards at fixed locations, where the work is undertaken. It is good practice to confirm the appointment to the individuals in writing as well, so that there is no confusion about the work expected of

them. RPSs should also be provided with sufficient information about their role (see paragraph 237).

Consultation with the RPA

295 The employer in control should consider the need for advice from the RPA about the suitability of RPS appointments.

Additional requirements for designated areas

(1)   Every employer who designates any area as a controlled or supervised area shall ensure that any such designated area is adequately described in local rules and that -

(a)   in the case of any controlled area -

(i)   the area is physically demarcated or, where this is not reasonably practicable, delineated by some other suitable means; and

(ii)   suitable and sufficient signs are displayed in suitable positions indicating that the area is a controlled area, the nature of the radiation sources in that area and the risks arising from such sources; and

(b)   in the case of any supervised area, suitable and sufficient signs giving warning of the supervised area are displayed, where appropriate, in suitable positions indicating the nature of the radiation sources and the risks arising from such sources.

Description of controlled and supervised areas

296 The responsibility for describing designated areas in local rules rests with the employer who designated the areas. Contractors undertaking work with ionising radiation within areas designated by other employers are not required to include a description in their local rules.

297 Where it is not appropriate to provide local rules for a particular supervised area (see regulation 17(1)), the extent of that area could be described either in the local rules of an adjacent controlled area or, where no such area has been designated, in the centrally held copy of the local rules for the site.

298 Designated areas will usually be described by reference to fixed features such as walls. Where the source of ionising radiation is mobile, the area(s) may be described generically, for example by reference to distances from the source or, if necessary, to distances from other objects irradiated by the source.

Physical demarcation of controlled areas

299 The main purpose of physically demarcating a controlled area is to help restrict unauthorised access. In determining whether suitable means of restricting access have been provided, the employer should consider the nature of the work and the likelihood that the means provided will restrict access to those people who are permitted to enter the area (see paragraphs 311-315). To be effective, the method of demarcation should clearly indicate the extent of the controlled area, with no possibility for doubt.

300 In most cases it will be appropriate to use physical features, for example existing walls and doors. Employers should provide temporary barriers where these physical features cannot be used; these barriers will usually need to be supervised. In such cases, the areas must be clearly delineated by other suitable means so employees (and other people as necessary) are aware that these areas exist.

301 Examples of situations where it may not be reasonably practicable to demarcate a controlled area are when:

(a) a vehicle transporting a radioactive substance is stationary at the side of a road because of breakdown and there is free-flowing traffic along the road;

(b) the area is an upper room of a multi-storey building and extends outside a window to which there is no access - the area would only be demarcated inside the building;

(c) a person has been administered with a radioactive substance as part of a medical exposure and is subsequently located in part of a room where the whole of that room has not been designated. Suitable means for delineation in this case would be a description, kept in a convenient place, of the extent of the controlled area around that patient; or

(d) the conditions requiring a controlled area arise from the use of X-ray equipment for dental radiography or veterinary radiography. The operator should be able to see any person in the vicinity of the controlled area and quickly de-energise the X-ray equipment from the normal operating position.

302 The RPA could be asked to advise on whether a barrier, or some other form of delineation, is necessary in the cases outlined in paragraph 301, for example to protect certain groups of individuals such as building workers and window cleaners.

303 Also, if work in an area is of short duration and transient in nature, it might not be reasonably practicable for the employer to demarcate the controlled area with barriers. An example is the use of a mobile X-ray set in a hospital ward. In such cases, the person operating the equipment (or other suitable person such as the RPS) should be able to restrict access to the area by continuous supervision. The extent of the designated area should be clearly described in the local rules. The operator will need to see all the boundaries of the area and either prevent unauthorised access or terminate the radiation exposure if someone tries to enter the controlled area. The presence of a controlled area should be clearly signalled to people in the vicinity. Only continuous supervision of the boundaries of an area is likely to be effective in preventing access by people where barriers cannot be provided.

Demarcation of an area after an incident

304 If a radiation accident or other incident occurs which involves a radiation source, it might take some time to fully demarcate a controlled area around that source. If that accident was foreseeable, the employer should have identified the need to restrict access to the affected area and taken reasonable steps for this to be done, as a component of the contingency plan prepared under regulation 12.

Warning signs for designated areas

305 Suitable warning signs are required for each designated controlled area. However, in the first case mentioned in paragraph 301 existing warning signs on the vehicle or packages will generally be sufficient. Suitable positions are likely to be at each entrance to the area or, in the case of temporary barriers, at frequent intervals where they can be seen by people approaching the barrier.

306 In addition, warning signs will be appropriate for some supervised areas. Where the extent of the supervised area is clearly set out in the local rules and the extent of the area is well understood by those who work there, it might not be appropriate to provide warning signs.

307 All warning signs should comply with the minimum requirements set out in Parts I-VII of Schedule 1 of the Health and Safety (Safety Signs and Signals) Regulations 1996.[13] Employers may add any supplementary text or cautionary notice they wish to the pictogram to make the sign appropriate to their situation. Signs should give sufficient information to alert employees to the risks arising from the source (eg X-rays or risk of inhaling or ingesting radioactive contamination). This should enable employees who have received appropriate training or instruction to know what action to take on entering the area, for example to wear personal protective equipment.

308 If it is not reasonably practicable to demarcate a controlled area it may not be practicable to provide warning signs. If so, one or more people should be instructed to attend to give a suitable verbal warning to anyone approaching the boundary of the controlled area.

Control over entry into areas

   (2)   The employer who has designated an area as a controlled area shall not permit any employee or other person to enter or remain in a such an area unless that employee or other person -

   (a)   being a person other than an outside worker, is a classified person;

   (b)   being an outside worker, is a classified person in respect of whom the employer has taken all reasonable steps to ensure that the person -

      (i)   is subject to individual dose assessment pursuant to regulation 21;

      (ii)   has been provided with and has been trained to use any personal protective equipment that may be necessary pursuant to regulation 8(2)(c);

      (iii)   has received any specific training required pursuant to regulation 14; and

      (iv)   has been certified fit for the work with ionising radiation which he is to carry out pursuant to regulation 24; or

   (c)   not being a classified person, enters or remains in the area in accordance with suitable written arrangements for the purpose of ensuring that -

      (i)   in the case of an employee aged 18 years or over, he does not receive in any calendar year a cumulative dose of ionising radiation which would require that employee to be designated as a classified person; or

75

(ii)   in the case of any other person, he does not receive in any calendar year a dose of ionising radiation exceeding any relevant dose limit.

Responsibility for control of access

309   The responsibility for controlling access to the area rests with the employer who designated that area under regulation 16. If the employer in control of a site hands the control of a particular area to another employer, for example a contractor undertaking maintenance work in that area, the second employer will be responsible for controlling access. Different requirements apply to access by classified persons and non-classified persons. Furthermore, there are additional requirements concerning access by classified persons who are outside workers. Any employee who enters a controlled area will require appropriate instruction under regulation 14.

310   In many cases, physical barriers and warning signs may be sufficient to restrict access to a controlled area when backed up by appropriate training and instruction in accordance with regulation 14. However, in some cases the employer in control may need to arrange for supervision of access points into the area to ensure that appropriate checks can be made.

Entry of classified persons

311   Under regulations 14, 21 and 24, employers of classified persons should ensure that any of their classified employees who enter controlled areas have received training appropriate to the area, suitable dose monitoring and adequate medical surveillance. If a classified person only has experience of work involving exposure to external radiation, the employer will need to consider what additional steps are required before permitting that person access to controlled areas designated on the basis of radioactive contamination

Entry of outside workers

312   Before an outside worker is allowed to enter the controlled area, the employer in control of that area needs to check that the person: has received the necessary training; is considered by the appointed doctor (or employment medical adviser) fit to undertake the work with ionising radiation; and is subject to routine dose assessment by an approved dosimetry service (see regulations 14, 21 and 24). Most of the information required, except for training details, should be found in the individual's radiation passbook (see paragraphs 409-413). The employer in control of the area will need to make arrangements for checking such passbooks before outside workers begin work in any controlled area on the site and for making estimates of the doses received (regulation 18(4)(a)). Both employers will have to co-operate to make sure that any outside workers sent to the site have had sufficient training to allow them to work safely in the controlled area and are provided with any necessary personal protective equipment (see paragraphs 114-118 and 163-164). This co-operation may best be arranged as part of the contractual agreement between the two employers.

313   In the absence of any formal agreement, the employer in control of the area will need to ask outside workers about the training they have received. If that training appears to be insufficient for the intended work then additional, specific training may be necessary before the employer allows them to start work in the area.

Entry of non-classified employees

314   Non-classified persons should only be allowed conditional access to controlled areas. The employer in control of the area should set out these

conditions in the written arrangements for entry into the area. These conditions should be aimed at ensuring adequate restriction of their exposure to ionising radiation and may include close supervision, the use of PPE and restrictions on the type of work done or the time spent in the area.

315 Non-classified persons may require access to a controlled area, for example where:

(a)   a whole room has been designated as a controlled area but the work with ionising radiation takes place in only one small area, such as a bench or fume cupboard;

(b)   a person enters the area for a limited period in order to carry out a simple maintenance job, for example to repair a central heating radiator, to witness a test or to carry out an inspection; and

(c)   work with ionising radiation is carried on intermittently in the area.

Written arrangements for non-classified employees

316 Written arrangements should be aimed at restricting the exposure of non-classified persons (as required by regulation 8). Only in exceptional cases would it be reasonable for a person working under such arrangements to receive a dose which approaches the levels specified in regulation 18(2)(c). The requirements under regulation 8(5) concerning women who are pregnant or breast-feeding may also be relevant. It may be sufficient to require such women to inform the line manager for the area and to restrict the type of work that they can do in the area. The written arrangements should also include provision for estimating the dose likely to be received (see paragraphs 319 and 320).

Entry by members of the public and employees who do not normally work with ionising radiation

317 Arrangements may have to be made to allow other employees or members of the public to enter a controlled area, for example for visitors to hospitals and organised parties visiting licensed nuclear installations (but see paragraph 318 for advice about members of the public who are 'comforters and carers'). The employer may also provide suitable dosemeters for these visitors, which would be one means of satisfying the requirements of regulation 18(3).

Entry of patients and comforters and carers

318 The requirement to restrict entry to controlled areas does not apply to any patient or research subject entering that area for the purpose of receiving a medical exposure (see regulation 3). However, a hospital may have to provide written arrangements to enable comforters and carers to enter a room designated as a controlled area in support of a relative or friend who has received an administration of a radiopharmaceutical for a therapeutic purpose. The purpose of these arrangements would be to restrict unnecessary exposure of such people; they are not subject to any dose limit. The arrangements will need to be applied sensitively in each case, taking full account of the informed wishes of the individuals concerned (see paragraphs 130 and 131).

Dose monitoring for non-classified persons

(3) An employer who has designated an area as a controlled area shall not permit a person to enter or remain in such area in accordance with the written arrangements under paragraph (2)(c), unless he can demonstrate, by personal dose monitoring or other suitable measurements, that the doses are restricted in accordance with that sub-paragraph.

319   Where the employer provides personal dosimetry it is desirable, but not essential, that dose monitoring for these individuals is carried out by an appropriate dosimetry service (ADS) approved under regulation 35. An ADS has to satisfy specific criteria to ensure that they operate to a satisfactory standard and are suitable for approval. Any employer who decides to make alternative arrangements for assessing doses to non-classified workers should be able to demonstrate that the measurements and assessments have been made to a satisfactory standard. Dosemeters or monitoring instruments (except those supplied by an ADS) used to demonstrate compliance with regulation 18(2)(b) will need to be subject to adequate tests, periodically, to ensure that any measurements made remain suitable.

320   Responsible employers would normally check the estimates of doses received by non-classified persons entering or working in the controlled area, to make sure the written arrangements are effective. It is advisable for employees who spend a significant amount of their time working in designated controlled areas to receive personal dose monitoring. Nevertheless, other types of measurement may be suitable to demonstrate compliance with regulation 18(2). In some cases, it may be appropriate to issue a personal dosemeter (eg a direct reading dosemeter) to one person in a group of visitors which stay together. The measurements made should be representative of the doses received by each member of the group. In general, personal dose monitoring should not be necessary for occasional external visitors who enter a controlled area, since the conditions for entry would normally ensure that the doses they receive will be very small. However, if appropriate, employees may need to carry a suitable dosemeter or other device for the purposes of the contingency plan (see regulation 12). When personal dosemeters are not used, a suitable measurement for the purposes of regulation 18(7) may be a measurement of the dose rate (or airborne contamination level) together with a record of the time spent in the controlled area. However, such measurements are only suitable if the dose rate (or contamination level) is known and is fairly constant or is known not to exceed a particular value.

(4) An employer who has designated an area as a controlled area shall, in relation to an outside worker, ensure that -

(a)   the outside worker is subject to arrangements for estimating the dose of ionising radiation he receives whilst in the controlled area;

(b)   as soon as is reasonably practicable after the services carried out by that outside worker in that controlled area are completed, an estimate of the dose received by that worker is entered into his radiation passbook; and

(c)   when the radiation passbook of the outside worker is in the possession of that employer, the passbook is made available to that worker upon request.

Providing and recording dose estimates for outside workers

321 Preferably, the estimate should be made using a suitable personal dosemeter (or personal air sampler in the case of internal radiation). The employer in control of the area needs to make a quick estimate of the total dose received, so this can be entered in the radiation passbook before the individual leaves the site once the services have been completed. Normally, it should be possible to make an entry in the passbook before the outside worker leaves the site, at least for external radiation. A quick dose estimate that tends to overstate the dose is preferable to a more accurate estimate that cannot be made for some days or even weeks.

322 The accuracy of the estimate and complexity of the method used should be commensurate with the magnitude of the expected dose; gross overestimates may result in unnecessary restrictions being placed on workers.

323 The employer in control of the area will need suitable arrangements for recording in the outside worker's radiation passbook the estimate of total dose received during the services. An estimate may be entered in the passbook at the controlled area itself, or on exit from the controlled area on completion of the work. Alternatively, the passbook may be lodged elsewhere on site and completed on production of a signed record of the cumulative dose estimate originating from the work area. Employees who record such estimates in radiation passbooks will need suitable training (see regulation 14).

324 The period covered by the estimate will often be different from the dose assessment period used by the approved dosimetry service for that person. The dose assessment will provide a more accurate picture of the dose received over a one-month or three-month period.

325 The dose estimate should be made and entered in the passbook when the services performed on that site by the outside worker have ended, or when workers leave the site in order to perform services for another employer in a controlled area on a different site, whichever is the sooner. Estimates do not have to be entered in passbooks each time outside workers physically leave the site (eg at the end of a day's work).

326 Where the work is to be done outside normal working hours, the employer in control may need to make special arrangements for entering the dose estimate in the passbook. If the estimate cannot be entered in the passbook while the outside worker is on site, the person's employer should be advised of the estimate as soon as practicable. If an estimate cannot be made because of an ongoing investigation into the exposure of the worker, the employer in charge is advised to note this fact in the outside worker's radiation passbook.

Providing information to non-classified persons

(5) The employer who carries out the monitoring or measurements pursuant to paragraph (3) shall keep the results of the monitoring or measurements referred to in that paragraph for a period of two years from the date they were recorded and shall, at the request of the person to whom the monitoring or measurements relate and on reasonable notice being given make the results available to that person.

Significant risk of spreading contamination

(6)   In any case where there is a significant risk of the spread of radioactive contamination from a controlled area, the employer who has designated that area as a controlled area shall make adequate arrangements to restrict, so far as is reasonably practicable, the spread of such contamination.

327   This provision only relates to the risk of contamination (see definition in regulation 2(1)) spreading outside controlled areas (eg from handling radioactive substances not in sealed form) leading to possible exposure to internal radiation. The risk is likely to be significant if:

(a)   the employer requires employees who enter the area to wear personal protective equipment to protect them against surface or airborne contamination; or

(b)   the spread of contamination would cause the employer to extend the restrictions on entry to encompass a wider area, or to require the use of particular procedures (including the use of personal protective equipment) outside the designated area.

An employer responsible for the controlled area will wish to have confidence that undesignated zones around such controlled areas, particularly where there is public access, remain essentially free from the presence of radioactive material.

Specific measures to prevent the spread of contamination

(7)   Without prejudice to the generality of paragraph (6), the arrangements required by that paragraph shall, where appropriate, include –

(a)   the provision of suitable and sufficient washing and changing facilities for persons who enter or leave any controlled or supervised area;

(b)   the proper maintenance of such washing and changing facilities;

(c)   the prohibition of eating, drinking or smoking or similar activity likely to result in the ingestion of a radioactive substance by any employee in a controlled area; and

(d)   the means for monitoring for contamination any person, article or goods leaving a controlled area.

328   This requirement is in addition to the requirements in the Workplace (Health, Safety and Welfare) Regulations 1992[26] for providing washing facilities, accommodation for clothing etc. The purpose of this regulation is to ensure that any radioactive contamination is not ingested or spread to areas outside the controlled area.

Washing facilities

329   The employer in control of the area should consider likely levels of contamination arising in controlled areas, in order to assess the type and extent of washing and changing facilities needed. The possibility of mishaps, such as accidental spillages, should also be taken into account in deciding whether it is appropriate to provide these facilities. The provision of washing facilities will be necessary for places where contamination is likely. What is adequate will vary from normal washing facilities, where only low levels are expected, to showers where high levels can be expected. Normally, the best position for these facilities will be next to the changing facilities.

330 Where washing facilities are appropriate, they would typically include at least wash-basins with hot and cold water supplied via jets or sprays which can be operated without using hands (eg foot or elbow operated). The employer will also need to provide soap and drying facilities such as disposable towels; nail brushes may also be needed. Static or roller towels will not be suitable in most situations.

331 Normally, the washing facilities will need to be conveniently accessible but situated so that they do not themselves become contaminated.

Changing facilities

332 Changing facilities are likely to be appropriate where protective clothing (other than disposable gloves) has to be worn in the area (see regulation 9(3)). The employer will need to provide a system to allow any protective clothing or respiratory protective equipment that has been worn or used, or any other clothing that may have been contaminated, to be left initially in the controlled area. The arrangements should prevent the spread of contamination from protective clothing to personal clothing.

333 Where protective clothing is used, or respiratory protective equipment is worn for work in a controlled area, it will normally be appropriate to provide the following at or outside the entrance to the area:

(a) a bench or barrier to demarcate the exit, designed so that any protective clothing and respiratory protective equipment which may be contaminated can be removed and left within the area;

(b) lockers on the 'clean' side of the bench or barrier where each person can leave uncontaminated clothing, shoes, etc;

(c) containers on the 'active' side of the barrier for discarded contaminated clothing; and

(d) a supply of clean protective clothing if not provided elsewhere.

334 In laboratories handling small quantities of radioactive substances, less onerous arrangements may be sufficient, provided that cross-contamination of clean and contaminated clothes and shoes is effectively prevented (but see regulation 9(3)).

Maintenance of washing and changing facilities

335 The term 'maintained' is defined in regulation 2(1). The facilities would not be in an efficient state unless they are maintained in a clean condition to prevent the build-up of contamination.

Eating, drinking and smoking

336 Where there is a significant risk of ingesting radioactive materials within the controlled area, because of surface contamination, the employer should make arrangements to prohibit eating, chewing, drinking or smoking etc in the area. However, employees can drink from a drinking fountain located in the area if it is constructed so that the water cannot be contaminated. Employers may wish to use the local rules, or signs posted in the area, to reinforce any prohibition.

337 Where it is necessary to prohibit eating or drinking, suitable facilities may have to be set aside for these activities to be carried out in an uncontaminated area.

Guidance

18(7)

Monitoring facilities

338 In general, it is appropriate to monitor for contamination wherever areas are designated as controlled because of the presence of unsealed radioactive substances, except where the nature of the substances makes such monitoring impracticable (eg radioactive gases). Where monitoring is undertaken, the employer in control of the area will need to decide whether portable monitors are sufficient or if installed personnel contamination monitors (eg hand and foot monitors) are required. The decision will depend on the nature of the work, the type of contamination being measured and the throughput of people and articles in the area. The monitors should be suitable for detecting significant contamination from the radionuclides present in the area. Such monitors will need to be properly maintained and tested to ensure they remain in effective working order. Advice from the RPA may be necessary (see regulation 13 and Schedule 5).

Regulation 19

Monitoring of designated areas

Regulation

19(1)

(1)   Every employer who designates an area as a controlled or supervised area shall take such steps as are necessary (otherwise than by use of assessed doses of individuals), having regard to the nature and extent of the risks resulting from exposure to ionising radiation, to ensure that levels of ionising radiation are adequately monitored for each such area and that working conditions in those areas are kept under review.

ACOP

19(1)

339 For areas designated on the basis of external radiation, adequate monitoring should include measurement of dose rates (averaged over a suitable period if necessary). For areas designated on the basis of internal radiation, adequate monitoring should include measurements of air activity and surface contamination where appropriate, taking into account the physical and chemical states of the radioactive contamination. In either case, the monitoring should be sufficient to indicate whether levels of radiation and contamination are satisfactory for continuing work with ionising radiation.

340 Monitoring should be designed to indicate breakdowns in controls or systems and to detect changes in radiation or contamination levels. In order to check the continued correct designation of areas, monitoring will be necessary both inside and outside the boundaries of controlled and supervised areas.

341 Employees carrying out the monitoring should be familiar with the proper use of the instruments and know how to interpret and record the results correctly.

Guidance

19(1)

Responsibility for monitoring designated areas

342 The responsibility for monitoring each controlled area and supervised area rests with the employer in control of that area. That employer will have responsibility for deciding to designate the area under regulation 16 (see paragraph 250).

Purpose of monitoring

343 The main purposes of monitoring are to:

(a) check that areas have been (and remain) correctly designated;

(b) help determine radiation levels and contamination from particular operations, so that appropriate control measures for restricting exposure can be proposed;

(c) detect breakdowns in controls or systems, so as to indicate whether conditions are satisfactory for continuing work in that area; and

(d) provide information on which to base estimates of personal dose for non-classified persons (see regulation 18(3)), outside workers (see regulation 18(4)) and classified persons for whom a dose assessment could not be made by the ADS (see regulation 22).

Adequate monitoring

344 In order to establish whether adequate monitoring is being achieved, the employer will need to consider:

(a) what kinds of measurements should be made (eg dose rates, surface contamination, air concentrations);

(b) the nature and quality of the radiation and the physical and chemical state of any radioactive contamination likely to be in the area;

(c) where the measurements should be made;

(d) how frequently, or on what occasions, these measurements should be made (including measurements forming part of contingency arrangements);

(e) what method of measurement would be used, for example direct measurement with an instrument, collection of air samples, use of wipe samples and what type of instrument to use (see paragraphs 349 and 350);

(f) who should carry out the measurements and what training they need (see guidance to regulation 14);

(g) how to ensure that the equipment continues to function correctly (see paragraphs 351-354 and regulation 19(2);

(h) what records should be kept (see paragraphs 362-366);

(i) how to interpret and review the results;

(j) the selection of reference levels and the action to be taken if they are exceeded; and

(k) when the monitoring procedures should be reviewed.

Role of the Radiation Protection Adviser

345 The radiation employer must consult the RPA about the implementation of requirements for controlled areas and supervised areas (regulation 13 and Schedule 5). These requirements include:

(a) the type and extent of the monitoring programme; and

(b) the regular calibration and checking of monitoring equipment to ensure it is serviceable and correctly used.

346 For this purpose, calibration is used in the general sense of establishing by measurement the response of the monitoring instrument to known radiation fields. The RPA should be able to advise the employer about the relevant radiation fields likely to be encountered in the designated areas, which the employer needs to be able to monitor adequately. The RPA will also advise on the routine checks that are necessary to ensure that the instrument remains serviceable on a daily basis. However, the employer will need the services of a qualified person to supervise or perform the examinations and tests required on instruments used to monitor those fields in accordance with this regulation (see paragraphs 357-359).

Selection, maintenance and testing of equipment used for monitoring designated areas

(2) The employer upon whom a duty is imposed by paragraph (1) shall provide suitable and sufficient equipment for carrying out the monitoring required by that paragraph, which equipment shall -

(a) be properly maintained so that it remains fit for the purpose for which it was intended; and

(b) be adequately tested and examined at appropriate intervals.

347 **Monitoring instruments used for measuring external radiation should be suitable for the nature and quality of the radiation concerned. Instrumentation used for measurements of air activity and surface contamination should be suitable for the physical and chemical state of the radioactive materials present.**

348 **Monitoring equipment should normally be tested and thoroughly examined at least once every year.**

Provision of suitable monitoring equipment

349 The employer in control of the area will need an adequate number of suitable instruments available for monitoring. In deciding the number of instruments to provide, the employer should take into account that sometimes they may need to be sent away for testing or repair.

350 The suitability of monitoring equipment will depend on the type, nature, intensity and energy of the radiation that has to be monitored and the conditions of use of the equipment. This process of matching monitoring equipment to the working environment would normally be carried out on the basis of information provided by the supplier about its performance in the environment in which it will be used.

351 The RPA should be consulted on the suitability of the instruments required for the type and energy range of radiations which the employer needs to monitor (see paragraphs 345 and 346 for an explanation of the RPA's role). Paragraph 357 gives advice on the separate role of the qualified person.

Maintenance of instruments

352 'Maintained' is defined in regulation 2(1). The functioning of instruments will need to be checked routinely as part of their maintenance. Routine checks typically include battery checks, zeroing and tests for lack of response to ensure that the equipment is still functioning satisfactorily and has suffered no obvious damage.

Periodic examination and testing

353 The purpose of regular thorough examinations and tests is to ensure that the monitoring equipment is not damaged, has not lost its calibration and is suitable for the expected duration of use until it is next thoroughly examined and tested. The ACOP advice in paragraph 348 gives practical guidance on an appropriate interval between tests and examinations for equipment in regular use. This interval may not be appropriate in some cases. Also, it is recognised that there may occasionally be practical difficulties in ensuring that such tests and examinations are always carried out every 12 months, although this should normally be the aim. The employer will need to rectify any significant faults that are identified: it may be necessary to retest the calibration following any repair.

354 Most thorough examinations and tests are likely to be undertaken by organisations and individuals specialising in this work. However, there is no reason why employers should not do their own examinations and tests, provided they have a qualified person with the necessary expertise and facilities to carry out, or supervise, the tests (see regulation 19(3)). Proper tests of monitoring instruments make use of sources or equipment that ensure a known accuracy of calibration traceable to appropriate national standards. In many cases this will involve the use of sources whose activity has been determined to a known accuracy, for example one obtained from a laboratory accredited by the United Kingdom Accreditation Service, or the use of equipment that has been previously calibrated to known accuracy. Authoritative guidance is available from the National Physical Laboratory measurement good practice guide.[27]

(3) Equipment provided pursuant to paragraph (2) shall not be or remain suitable unless -

(a) the performance of the equipment has been established by adequate tests before it has first been used; and

(b) the tests and examinations carried out pursuant to paragraph (2) and sub-paragraph (a) above have been carried out by or under the supervision of a qualified person.

355 All instruments should be individually calibrated before first use and as part of the annual examination and test.

356 Qualified persons should possess the necessary expertise in instrumentation, theory and practice appropriate to the type of instrument to be tested.

Qualified persons for testing equipment

357　Qualified persons may be either employees of the employer who designated the area or of an instrument manufacturer, supplier or specialist test house. Qualified persons need to be fully conversant with, and have knowledge and understanding of, currently accepted testing standards and relevant technical guidance on testing the type of monitoring equipment under test (eg the National Physical Laboratory measurement good practice guide referred to in paragraph 354). They will often have responsibility for developing test protocols for the monitoring equipment, taking into account the intended use of the monitoring equipment as stated by the employer.

358　It is the duty of the employer in control of the area to determine what types of test are appropriate for the circumstances in which the equipment is used to monitor levels of ionising radiation arising from the work being undertaken and to advise the qualified person accordingly. The qualified person is the expert on the nature and frequency of the tests required for these instruments and is responsible for completing the tests, appropriate to the working environment specified by the employer, competently and comprehensively. The qualified person should ensure that their clients know exactly what sort of examination and calibrations they have contracted for and any relevant limitations (see paragraphs 345 and 346 for advice on the separate role of the RPA).

359　The employer is not obliged to appoint the qualified person formally in writing. However, it is important that both parties clearly understand the scope of the testing work required, for example as part of a contractual arrangement. Records of any tests undertaken to satisfy the requirements of regulation 19(4) must be authorised by the qualified person. These records should provide a traceable link to the qualified person responsible for the tests carried out on particular monitoring equipment.

Testing of monitoring equipment before first use

360　Testing of instruments before they are taken into use will usually comprise individual tests of each monitoring instrument and individual calibration. Where type testing has been carried out, selected tests are usually undertaken to verify that an instrument conforms to type. If no type test data are available, then before first use, tests are required to establish the instrument's performance for its intended use and also to determine any limitations that may render it unsuitable for certain applications. Information on the limitations of any equipment and its accuracy of calibration should be available to any person who uses, or may use, it.

361　Where (as part of the service to employers) manufacturers, suppliers and organisations specialising in monitoring equipment carry out the tests referred to in regulation 19(3), these tests should still be carried out by, or under the immediate supervision of, a qualified person. The qualified person will be able to decide on the testing required in the light of the type test data or any other reliable information they have about the design criteria and performance characteristics of that particular type of equipment, and of its intended use.

Monitoring and test records

(4)　The employer upon whom a duty is imposed by paragraph (1) shall -

(a)　make suitable records of the results of the monitoring carried out in accordance with paragraph (1) and of the tests carried out in accordance with paragraphs (2) and (3);

(b)   ensure that the records of the tests carried out pursuant to sub-paragraph (a) above are authorised by a qualified person; and

(c)   keep the records referred to in sub-paragraph (a) above, or copies thereof, for at least 2 years from the respective dates on which they were made.

362   Suitable monitoring records should include the date, time and place of monitoring and confirm that controlled and supervised areas are correctly designated and show where levels are being approached which may require investigatory or remedial action to be taken. For areas designated on the basis of external radiation there should be an indication of the nature and quality of the radiation in question. For areas designated on the basis of internal radiation the results should indicate the nature and physical and chemical states of radioactive contamination unless this is inappropriate.

363   Any records of instrument tests carried out for the purposes of regulations 19(2) and (3) should be signed by a qualified person. The name and contact details of that person should be stated in the record.

364   In addition to the details mentioned in the ACOP advice in paragraph 362, the record of the monitoring results would typically show the radiation level found (ie dose rate, air concentration or surface contamination) and the conditions existing, where relevant, in relation to the radiation level measured. However, there may be no need to retain many routine results of monitoring. The employer, in consultation with the RPA, should decide in advance what needs to be recorded (see regulation 13 and Schedule 5). The employer might also include the details of the person carrying out the monitoring and the particular equipment used. These details might be important because the employer will need to demonstrate that:

(a)   the person undertaking the monitoring had received adequate training as required by regulation 14; and

(b)   the equipment used for monitoring was suitable and adequately tested as required by regulation 19(2).

365   A suitable record of equipment monitoring tests might typically include: identification of the equipment; the calibration accuracy over its range of operation for the types of radiation that it is intended to monitor; the date of the test and the name and signature of the qualified person under whose direction the test was carried out (see ACOP advice in paragraph 363).

366   The qualified person takes responsibility for the accuracy of the information in the test certificate or other record of the test by signing it. However, employers have responsibility for ensuring that instruments are suitable for their working environments. The qualified person cannot be responsible for failing to test an instrument for an appropriate range of operation if the employer does not make it clear where the instrument will be used, ie the range of energies and types of radiation encountered in the working environment.

Regulation 20

Designation of classified persons

Regulation

20(1)-(2)

(1)   Subject to paragraph (2), the employer shall designate as classified persons those of his employees who are likely to receive an effective dose in excess of 6 mSv per year or an equivalent dose which exceeds three-tenths of any relevant dose limit and shall forthwith inform those employees that they have been so designated.

(2)   The employer shall not designate an employee as a classified person unless -

(a)   that employee is aged 18 years or over; and

(b)   an appointed doctor or employment medical adviser has certified in the health record that that employee is fit for the work with ionising radiation which he is to carry out.

ACOP

20(1)-(2)

Who needs to be a classified person

367   In deciding whether a person should be classified, the employer should take account of the potential for exposure to ionising radiation (including the possibility of accidents etc which are likely to occur) as a result of the work the individual is required to undertake.

368   An employer should designate as a classified person any employee who works with any source of ionising radiation which is capable of giving rise to a dose rate such that it is reasonably foreseeable an employee could receive an effective dose greater than 20 millisieverts or an equivalent dose in excess of a dose limit within several minutes.

Guidance

20(1)-(2)

369   In deciding if an employee needs to be a classified person the employer can take account of past doses received by the employee. However, it is not sufficient to rely on the individual's avoidance of such a dose in the past to argue that classification is unnecessary. In many cases, the reason an employer will designate an employee as a classified person will be that the individual undertakes much of their work in controlled areas. However, this fact alone is not sufficient to require designation, particularly where the work with ionising radiation in the area is intermittent, or carried on in one small part of the area, and in either case employees are unlikely to receive significant exposure. Where employees are not designated as classified persons, employers will need to provide suitable written arrangements to enable them to enter any controlled areas (see paragraph 316).

370   Employees should be informed as soon as possible if they have been designated as classified persons, so they are aware of their responsibilities to co-operate with the employer under regulation 34. Employers will have to consider arranging medical examinations for employees (see paragraphs 452 and 453) and specific training (see paragraphs 237 and 238) before assigning them to a post as classified persons.

New employees who were classified persons in their previous employment

371   Any employer who engages individuals to work as classified persons must ensure that an appointed doctor or employment medical adviser has certified that they are fit for the intended type of work before they start the work as a classified person (see paragraphs 452 and 453).

372 Normally, when a classified person changes employment, the new employer checks that the person has been certified fit for work in the last 12 months by obtaining a copy of that certificate from the previous employer. The employer will need to take account of any restrictions in that certificate (see paragraph 463). If the nature of the work with ionising radiation is significantly different from that covered by the certificate, the employer may need to arrange for the new employee to receive a medical examination from an appointed doctor or employment medical adviser (see regulation 24).

373 When steps have been taken to meet the requirements for medical surveillance the new employee can start work as soon as training/induction is complete. However, the employer will need to ensure that the radiation dose received is assessed (see regulation 21).

Employees on short-term contracts

374 Employers that take on workers who need to be designated as classified persons on short-term contracts will need to take special care to ensure that these employees are covered by arrangements for medical surveillance, dose assessment and dose record-keeping required by regulations 21 and 24.

Female employees

375 Employers should be aware of the special restrictions on working conditions for women who have declared themselves to be pregnant or who are breast-feeding (regulation 8(5)), and also of the additional dose limit for women of reproductive capacity (see Schedule 4 and paragraph 194).

When can an employee cease to be treated as a classified person?

(3)   The employer may cease to treat an employee as a classified person only at the end of a calendar year except where -

(a)   an appointed doctor or employment medical adviser so requires; or

(b)   the employee is no longer employed by the same employer in a capacity which is likely to result in significant exposure to ionising radiation during the remainder of the relevant calendar year.

376 **Exposure is significant if the employee is likely to receive an effective dose at a rate exceeding 1 millisievert per year as a result of work in the new post.**

377 Employers can only cease designation of employees as classified persons before the end of the calendar year in special circumstances, as follows:

(a)   an employee ceases employment (a termination record is required - see paragraph 397);

(b)   the appointed doctor (or employment medical adviser) certifies in the health record that an employee should not be engaged in work with ionising radiation as a classified person (see paragraph 463);

(c)   an employee is transferred by the employer to new duties which do not involve any significant exposure to ionising radiation, for example if the employee is promoted to a supervisory job which involves no further work in a designated area. If the new post did involve significant exposure, systematic monitoring of doses, as required by regulation 21, would need to continue and the person should still be treated as a classified person.

378 In most cases, employees will only cease to be classified persons if they cease employment with their current employer. Where the employer wishes to cease treating an employee as a classified person on transfer to new duties, that employer would have to be satisfied that the employee will not return to their original duties before the end of the calendar year or undertake any other work involving significant exposure. If it is likely that the individual will return to work involving significant exposure to ionising radiation within the year, for example because they work for a contractor undertaking work intermittently on licensed nuclear installations, he or she should normally continue to be treated as a classified person. However, the employer should advise the approved dosimetry service of periods when that employee does not receive any exposure to ionising radiation, as no dose assessments would be required for those periods (see regulation 22(4)).

379 Employers will have to reconsider the need to designate employees as classified persons if such individuals revert to their original duties in any subsequent calendar year.

380 Whether or not an employee remains as a classified person, the employer will have to ensure that doses received over any relevant dose limitation period do not exceed the limits set out in Schedule 4.

Dose assessment and recording

(1) Every employer shall ensure that -

(a) in respect of each of his employees who is designated as a classified person, an assessment is made of all doses of ionising radiation received by such employee which are likely to be significant; and

(b) such assessments are recorded.

(2) For the purposes of paragraph (1), the employer shall make suitable arrangements with one or more approved dosimetry service for -

(a) the making of systematic assessments of such doses by the use of suitable individual measurement for appropriate periods or, where individual measurement is inappropriate, by means of other suitable measurements; and

(b) the making and maintenance of dose records relating to each classified person.

(3) For the purposes of paragraph (2)(b), the arrangements that the employer makes with the approved dosimetry service shall include requirements for that service -

(a) to keep the records made and maintained pursuant to the arrangements or a copy thereof until the person to whom the record relates has or would have attained the age of 75 years but in any event for at least 50 years from when they were made;

(b) to provide the employer at appropriate intervals with suitable summaries of the dose records maintained in accordance with sub-paragraph (a) above;

(c) when required by the employer, to provide him with such copies of the dose record relating to any of his employees as the employer may require;

(d) when required by the employer, to make a record of the information concerning the dose assessment relating to a classified person who ceases to

be an employee of the employer, and to send that record to the Executive and a copy thereof to the employer forthwith, and a record so made is referred to in this regulation as a "termination record";

(e) *within 3 months, or such longer period as the Executive may agree, of the end of each calendar year to send to the Executive summaries of all current dose records relating to that year;*

(f) *when required by the Executive, to provide it with copies of any dose records;*

(g) *where a dose is estimated pursuant to regulation 22, to make an entry in a dose record and retain the summary of the information used to estimate that dose;*

(h) *where the employer employs an outside worker, to provide, where appropriate, a current radiation passbook in respect of that outside worker; and*

(i) *where the employer employs an outside worker who works in Northern Ireland or another member State, maintain a continuing record of the assessment of the dose received by that outside worker when working in such place.*

Engaging suitable approved dosimetry services

381 Employers who designate employees as classified persons under regulation 20(1) will need to engage approved dosimetry services to undertake any necessary dose assessments, to open and maintain dose records and to provide relevant information from those records.

382 Dosimetry services are approved by HSE (or a body specified by HSE) under regulation 35 for one or more of the following specific purposes:

(a) the measurement and assessment of whole-body or part-body doses arising from external radiation (notably X-rays, gamma rays, beta particles or neutrons);

(b) the assessment of doses from intakes of specified classes of radionuclides;

(c) the assessment of doses following an accident or other incident (see regulation 23); and

(d) the co-ordination of individual dose assessments by other approved services, making, maintaining and keeping dose records, and the provision of summary information.

An organisation may hold certificates of approval for more than one of these functions. However, that organisation will only undertake those approved functions covered by contractual or other formal arrangements with the employer.

383 The aim of approval is to ensure, as far as possible, that doses are assessed on the basis of accepted national standards and that dose records bring together all such dose assessments, helping employers check that doses are being kept as low as reasonably practicable and dose limits are not exceeded. For this purpose, HSE has published criteria for approval which must be met by a dosimetry service seeking approval or wishing to remain approved.

384 The certificate of approval issued to each ADS shows the purpose for which it has been approved and any limitations on that approval. Employers should consider consulting the RPA about the types of dosimetry service required for the assessment and recording of doses received by the classified persons. They should then engage sufficient and suitable ADS for those purposes.

Significant doses

385 When a person is subject to the requirements for dose assessment, all doses likely to be significant are to be assessed (except those received as a result of natural background radiation at normal levels and the person's own medical exposure). Therefore, if personal dosemeters are used for assessing a particular component of dose, they should be worn at all times while at work when there is likely to be significant occupational exposure from that component.

386 If dose assessments are made for exposure to external gamma radiation or X-rays, the employer may have to decide whether additional components of dose should be assessed as well. These components might include, for example, dose from neutrons or committed dose from internal radiation. The employer's decision will depend on the expected magnitude and variations in that component. Normally, effective doses from an additional component which can reasonably be expected to exceed about 1 millisievert a year would be regarded as significant in this context for that component of dose. However, it may not always be appropriate to assess a particular component of dose at this level by individual measurement.

Methods of assessing individual dose

387 There are various methods of assessing personal dose. The choice of method will depend on the circumstances of individual cases, including the nature of the work and the type of ionising radiation which is to be measured. Examples of different methods are the use of different types of personal dosemeters to assess: the whole-body dose and skin dose from external radiation; the dose to the skin of the hands; and the dose to the whole body from neutron radiation.

388 In many cases, employers will only need an assessment to be made of doses received by their employees from external X-rays, gamma radiation and beta particles, using an appropriate type of personal dosemeter. Assessment of committed doses arising from intakes of radionuclides into the body (by inhalation, ingestion or via wounds) is generally more difficult and may involve a combination of techniques. These include: personal air samplers; biological samples (particularly urine); and monitoring of part of the body (eg the thyroid in the case of radioactive iodine) or the whole body with sensitive detectors.

Appropriate dose assessment periods

389 Often the dose assessment period (eg the issue period for dosemeters) will be one month, but periods as long as three months may be appropriate where doses are very low. For external radiation, the choice of assessment period will depend on the dose rates to which people are exposed, the ability of the dosemeter to measure low doses, the magnitude of the expected dose and the stability of the stored image or signal on the dosemeter over time. More frequent assessment will be appropriate where there is a significant risk of accidental exposure, for example when working with large sealed sources outside a shielded enclosure. For particular types of dosemeter the issue period might be as short as one shift. The longer the assessment period the

more difficult it becomes to determine when (and why) an individual's dosemeter has received an unusually high dose in the event of an accident or other incident.

Role of the employer

390 The employer is responsible for putting the arrangements with the ADS into effect. Arrangements may be needed for: handing out dosemeters, air samplers etc; the secure storage of dosemeters or samples; and the collection and despatch of dosemeters or samples to the ADS. The employer may also need adequate arrangements to protect dosemeters from accidental exposure when not being worn, for example by avoiding the use of X-ray security machines for dosemeters sent to the employer by the ADS. The employer may find it helpful to consult relevant appointed safety representative(s) on the general arrangements for the issuing, collecting and secure storage of dosemeters. Further advice is given in the HSE information sheet on dose assessment and recording.[28]

Keeping classified employees informed about the arrangements

391 The employer should ensure that classified persons have adequate training, information and instruction to enable them to adhere to the administrative arrangements made for the assessment and recording of the doses they receive (regulation 14). In particular, such people need to be made aware of their duties under regulation 34(4) to comply with reasonable requirements for the purpose of making measurements and assessments of the doses they receive, and of their right to see copies of dose summaries etc as provided under regulation 21(6).

392 The Data Protection Act 1998[29] requires that data subjects, classified persons in this case, are made aware of the nature of the information held on them. The information provided to classified persons under IRR99 might include details of the type of data held by the ADS and HSE's Central Index of Dose Information (CIDI),[30] as provided by this regulation.

Keeping dose records

393 Regulation 21(3) requires a formal dose record to be maintained for each classified person by a dosimetry service approved for that specific purpose. This will ensure that all assessed doses are properly recorded and summed for the calendar year (or other dose limitation period). The ADS responsible for maintaining the dose record will also help to ensure that a dose history is obtained for the person, either from the previous ADS or from HSE's CIDI.[30]

394 Occasionally, employers may receive reports of doses to their employees directly from the ADS which assessed those doses. This situation is only likely to arise where the ADS is unsure whether these employees are classified persons. If the employees are classified persons, the employer should pass these reports of doses to the dosimetry service approved for record-keeping - only the records kept by this service would meet the requirement under regulation 21(3).

Dose summaries

395 The main purpose of dose summaries is to help the employer check that doses are being adequately restricted. For example, they will reveal significant differences between doses received by individuals undertaking similar work. Typically, dose summaries will be provided at the same frequency as dose assessment periods. Employers will want to ensure that any arrangements

made with an ADS for providing dose summaries suit their particular circumstances. Intervals greater than three months are unlikely to be suitable (see paragraph 389).

396 The employer will also want to be informed as soon as possible when the cumulative dose for a classified person is approaching, or has exceeded, a relevant dose limit. The arrangements made with the ADS engaged to maintain dose records would normally provide for this prompt information to the employer.

Termination records

397 The main purpose of termination records is to provide the new employer with relevant dose information when a classified person changes employment. It also provides a summary for the employer and the employee of the doses received during employment as a classified person with that employer. In some cases, an employee might cease to be a classified person but continue to be employed by the same employer in a different post for some years. The employer must ensure that a termination record has been issued when that employee changes employer or retires. In such cases, the employer may prefer the ADS to issue a termination record immediately. This will ensure that the task is not forgotten. Such an arrangement complies with regulations 21(3)(d) and 21(6)(b), except that a replacement termination record will be needed if that employee resumes work as a classified person before ceasing employment with the employer. In any event, the employee may ask the employer to provide a copy of his or her dose record under regulation 21(6)(a), on giving reasonable notice.

Provision of radiation passbooks for outside workers

398 The radiation passbook approved by HSE under these Regulations is a document personal to an outside worker that cannot be transferred to other employees. The outside worker can retain the passbook when transferring to employment with another employer. Employers may continue to use radiation passbooks approved under the Ionising Radiations (Outside Workers) Regulations 1993 (OWR93) which have been issued to their outside workers before 30 April 2000 (see regulation 39(4)) until those passbooks are full. For outside workers who start work, or transfer to different employers after that date, the employer will need to ensure that they have passbooks approved under the Ionising Radiations Regulations 1999.[1] New passbooks can only be obtained from an ADS approved for the co-ordination of dose assessments and maintenance of dose records. The ADS can only purchase passbooks from HSE. Where appropriate, employers of classified persons should arrange for their ADS to allocate passbooks on their behalf. Exceptionally, employers may ask the ADS to issue passbooks to the employer who will then individually allocate them to the outside workers. Passbooks approved under these Regulations can be transferred with an outside worker who changes employer and used by the new employer until full, provided relevant particulars such as the new employer's name and address are entered.

Continued dose assessment for outside workers

399 Regulation 21(3)(i) extends an employer's duty for arranging statutory dose assessment to cover services undertaken for another employer in Northern Ireland or another member State. Where the period of work outside Great Britain is less than a dose assessment period, this could be done by requiring the outside worker to continue wearing whatever personal dosemeter

94

is appropriate and/or normally worn, providing the outside worker will not receive any significant exposure to new components of dose (eg intakes of radionuclides, neutron dose or dose to the hands).

400 When the period of work is longer than a dose assessment period, the employer would generally arrange for the ADS to send replacement personal dosemeters to the outside worker as necessary and for these to be returned to the ADS for assessment. Alternatively, the employer may choose to make arrangements (for example by contract with the other employer) for the dose assessment(s) to be made and notified by a local dosimetry service that has been recognised for such purposes by the relevant competent authority in the member State (or where appropriate Northern Ireland). The arrangements made by the employer should ensure that any such dose assessment for an outside worker is included in that person's dose record in Great Britain.

Dose assessment and record-keeping for employees working entirely outside the UK

401 If employees only work with ionising radiations in another member State or elsewhere outside the UK there is no duty under these Regulations to make arrangements for dose assessment and maintenance of dose records for those individuals. Instead, the employer should comply with relevant national legislation, which in the case of other member States will implement the relevant requirements of the Basic Safety Standards Directive 96/29/Euratom.[3] Dose records for these individuals are clearly important but they should not be confused with records maintained for employees designated as classified persons for work in Great Britain. In particular, employers are not expected to make arrangements to report to HSE annual summaries for such employees under regulation 21(3)(e).

Information for the ADS

(4) The employer shall provide the approved dosimetry service with such information concerning his employees as is necessary for the approved dosimetry service to comply with the arrangements made for the purposes of paragraph (2).

402 The effectiveness of the arrangements with the ADS will depend on information provided by the employer about each classified person. The employer will need to provide accurate and timely information to the ADS responsible for co-ordination and dose record-keeping about new employees whom the employer intends to designate as classified persons. This will enable the ADS to establish a new dose record and to obtain a dose history for that person. The ADS will need to know such details as the employee's full name, national insurance number, date of birth, gender, date of starting work as a classified person with that employer and the type of work that person will be undertaking. The ADS will also need to know about any changes to these details.

403 The employer will also have to provide the ADS responsible for assessing doses with certain information, at the end of each dose assessment period. This may include, for example, details about which classified persons wore specified dosemeters for that dose assessment period. The ADS will also need to know whether any classified persons are suspected of receiving an accidental exposure (see paragraph 440).

404 Employers should inform the ADS responsible for record-keeping about any dose records or summaries of doses they hold for those employees they intend to designate as classified persons.

Individuals who have more than one employer concurrently

405 Where an employee has more than one employer at the same time (or undertakes work with ionising radiation as a self-employed person outside the normal working hours of his or her employer), each employer will need to try and establish the total dose received by the worker (see also regulation 15). This is necessary to establish that dose limits or investigation levels have not been exceeded. The ADS responsible for maintaining dose records for the relevant employer will need to be informed of any information the employer has on concurrent dose records established for the individual under arrangements made by another employer. It may be feasible for the two dosimetry services to co-operate on the dose information available for that employee.

Individuals on short-term contracts

406 Employers employing individuals on short-term contracts as classified persons will need to make sure that the ADS responsible for opening and maintaining dose records has sufficient information about the doses received by those persons for the calendar year to date (see also paragraph 374 on regulation 20). Where the individual is already subject to a five-year dose limit under regulation 11(2) and Schedule 4, or needs to be made subject to this limit, information may well be required for the previous four years (excluding any years before the year 2000). The employer may obtain that information from a copy of the termination record provided to the individual by the former employer. Failing that, the employer's ADS may need to exchange information with the former ADS. HSE's CIDI[30] maintains information related to classified persons showing which ADS is currently responsible for maintaining dose records.

407 In each case, the results of dose assessments for these short-term classified persons need to be entered into a dose record held for that person by an ADS, even when the employee's contract is shorter than the usual dose assessment period. Normally, the employer will register any new employee with the ADS for dose record-keeping before returning any personal dosemeters worn by that person to the ADS for assessment.

408 Employers will need to take action along the lines discussed in paragraph 407 even if the same employee is taken on repeatedly for a series of short but separate contacts.

Keeping outside worker's radiation passbooks up to date

(5) An employer shall -

(a) ensure that each outside worker employed by him is provided with a current individual radiation passbook which shall not be transferable to any other worker and in which shall be entered the particulars set out in Schedule 6; and

(b) make suitable arrangements to ensure that the particulars entered in the radiation passbook are kept up-to-date during the continuance of the employment of the outside worker by that employer.

409 **Entries in passbooks should only be made by people who have been authorised by the approved dosimetry services or the appropriate employer to make such entries. Suitable arrangement should include written instructions, specifying who does what and when, unless this would clearly be inappropriate in the circumstances.**

410 Employers have to make sure their outside workers have valid radiation passbooks and that the information in these passbooks is kept up to date. Passbooks cannot be transferred between employees but they may continue to be used when a classified person changes employer. In such cases the new employer should enter the relevant details, including the employer's name and address, in the passbook. If the classified person has left the radiation passbook with the previous employer, or the passbook is full, the new employer will need to obtain a fresh passbook from the ADS responsible for record-keeping. Regulation 39(4) allows employers to continue using passbooks approved under OWR93 for existing outside workers until the passbooks are full (see paragraph 398).

411 If the employer is based in a member State which does not provide an authorised passbook, contracts may specify that the employer should obtain one from a dosimetry service approved in Great Britain for record-keeping. The employer will be responsible for ensuring it contains the necessary information.

412 As explained in paragraph 410 radiation passbooks may be used until they are full. However, closing passbooks at the end of a calendar year before they are full may avoid the need to transfer cumulative dose information for a part-year to any replacement passbook.

413 The employer of the outside worker should generally make sure that the details in the passbook - such as the date and result of the last medical review and the cumulative dose assessment for the year so far - are brought up to date as far as possible before the outside worker is to undertake new services in the controlled area of another employer.

21(5)

Providing dose information to classified persons

(6) The employer shall -

(a) at the request of a classified person employed by him (or of a person formerly employed by him as a classified person) and on reasonable notice being given, obtain (where necessary) from the approved dosimetry service and make available to that person -

(i) a copy of the dose summary provided for the purpose of paragraph (3)(b) relating to that person and made within a period of 2 years preceding the request; and

(ii) a copy of the dose record of that person; and

(b) when a classified person ceases to be employed by the employer, take all reasonable steps to provide to that person a copy of his termination record.

21(6)

414 This provision allows classified persons to obtain personal dose monitoring information from the employer. Many employers will provide copies of dose summaries, or an extract of such information, to their employees without receiving a request, in order to show their commitment to keeping doses as low as reasonably practicable.

21(6)

(7) The employer shall keep a copy of the summary of the dose record received from the approved dosimetry service for at least 2 years from the end of the calendar year to which the summary relates.

21(7)

Estimated doses and special entries

(1) Where a dosemeter or other device is used to make any individual measurement under regulation 21(2) and that dosemeter or device is lost, damaged or destroyed or it is not practicable to assess the dose received by a classified person over any period, the employer shall make an adequate investigation of the circumstances of the case with a view to estimating the dose received by that person during that period and either -

(a) in a case where there is adequate information to estimate the dose received by that person, shall send to the approved dosimetry service an adequate summary of the information used to estimate that dose and shall arrange for the approved dosimetry service to enter the estimated dose in the dose record of that person; or

(b) in a case where there is inadequate information to estimate the dose received by the classified person, shall arrange for the approved dosimetry service to enter a notional dose in the dose record of that person which shall be the proportion of the total annual dose limit for the relevant period,

and in either case the employer shall take reasonable steps to inform the classified person of that entry and arrange for the approved dosimetry service to identify the entry in the dose record as an estimated dose or a notional dose as the case may be.

(2) The employer shall, at the request of the classified person (or a person formerly employed by that employer as a classified person) to whom the investigation made under paragraph (1) relates and on reasonable notice being given, make available to that person a copy of the summary sent to the approved dosimetry service under sub-paragraph (a) of paragraph (1).

22(1)-(2)

415 The employer's investigation should take account of the following where relevant:

(a) details of the pattern of work of the individual such as the time spent in particular controlled and supervised areas;

(b) measurements from any additional dosemeter or direct reading device worn by the person concerned;

(c) individual measurements made on other employees undertaking the same work with ionising radiations; and

(d) the results of monitoring for controlled and supervised areas carried out in accordance with regulation 19.

22(1)-(2)

Lost or destroyed dosemeters etc

416 Dosemeters used to assess individual dose for classified persons may occasionally be lost or destroyed. Also, it may not be possible to assess the dose received by a person because the dosemeter was severely damaged or exposed to ionising radiation when not being worn (eg in X-ray security machines). In these cases, employers need to advise the ADS responsible for the relevant dose assessment as soon as they are informed of the problem and undertake an investigation. The employer should generally consult the RPA about this investigation (see paragraph 217).

22(1)-(2)

Nature of the employer's investigation

417 The employer's investigation should be aimed at establishing both the cause of the loss or damage to the dosemeter, to prevent a recurrence, and an adequate estimate of the dose received by the individual. The employer will find it relatively easy to establish an estimate where the person was wearing an additional dosemeter or a direct reading device. In other cases, it may be possible to make an adequate estimate of the person's dose by considering information such as their pattern of work during the dose assessment period, together with any relevant monitoring data collected in accordance with regulation 19(1). If the dosemeter was damaged, or exposed when not being worn, the ADS responsible for routine dose assessments may be able to assist the investigation by examining and reporting on the dosemeter separately from those dosemeters returned by other employees. The employer should usually tell the ADS that an investigation is under way.

418 If the investigation provides adequate evidence for an estimate of the dose received by the classified person, the employer should arrange for the ADS responsible for maintaining the dose record to insert that estimate in the record. If it is not possible to make an adequate estimate of the dose received, the employer should arrange for the ADS to insert a 'notional dose' in the record, but a notional dose should only be used as a last resort. The employee concerned should also be informed.

Providing information to the approved dosimetry service

419 Where an estimate is made, the employer will also need to make an adequate summary of the evidence used to produce that estimate (for example monitoring data for designated areas), and provide both to the ADS responsible for maintaining the dose record.

Replacing doses recorded in dose records with special entries

(3)   Subject to paragraphs (5) and (8), where an employer has reasonable cause to believe that the dose received by a classified person is much greater or much less than that shown in the relevant entry of the dose record, he shall make an adequate investigation of the circumstances of the exposure of that person to ionising radiation and, if that investigation confirms his belief, the employer shall, where there is adequate information to estimate the dose received by the employee -

(a)   send to the approved dosimetry service an adequate summary of the information used to estimate that dose;

(b)   arrange for the approved dosimetry service to enter that estimated dose in the dose record of that person and for the approved dosimetry service to identify the estimated dose in the dose record as a special entry; and

(c)   notify the classified person accordingly.

(4)   The employer shall make a report of any investigation carried out under paragraph (3) and shall preserve a copy of that report for a period of 2 years from the date it was made.

(5)   Paragraph (3) shall not apply -

(a)   in respect of a classified person subject only to an annual dose limit, more than 12 months after the original entry was made in the record; and

(b)   in any other case, more than 5 years after the original entry was made in the record.

(6)   Where a classified person is aggrieved by a decision to replace a recorded dose by an estimated dose pursuant to paragraph (3) he may, by an application in writing to the Executive made within 3 months of the date on which he was notified of the decision, apply for that decision to be reviewed.

(7)   Where the Executive concludes (whether as a result of a review carried out pursuant to paragraph (6) or otherwise) that -

(a)   there is reasonable cause to believe the investigation carried out pursuant to paragraph (3) was inadequate; or

(b)   a reasonable estimated dose has not been established,

the employer shall, if so directed by the Executive, re-instate the original entry in the dose record.

(8)   The employer shall not, without the consent of the Executive, require the approved dosimetry service to enter an estimated dose in the dose record in any case where -

(a)   the cumulative recorded effective dose is 20mSv or more in one calendar year; or

(b)   the cumulative recorded equivalent dose for the calendar year exceeds a relevant dose limit.

420   An estimate of the dose received should be regarded as much greater than or much less than the original entry in the dose record for a particular period if:

(a)   the dose received differs from the original entry in the dose record by at least 1 millisievert for recorded doses of 1 millisievert or less; or

(b)   the dose received differs from the original entry in the dose record by a factor of 2 or more for recorded doses in excess of 1 millisievert but less than the relevant dose limit; or

(c)   the dose received differs from the original entry in the dose record by a factor of 1.5 or more for recorded doses above the relevant dose limit.

421   The employer's investigation into the circumstances of the exposure should take account of:

(a)   relevant information provided by the approved dosimetry service;

(b)   details of the pattern of work of the individual such as the time spent in particular controlled and supervised areas;

(c)   measurements from any additional dosemeter or direct reading device worn by the person concerned;

(d)   individual measurements made on other employees undertaking the same work with ionising radiations; and

(e) the results of monitoring for controlled and supervised areas carried out in accordance with regulation 19.

422 The information used to estimate the dose received will be adequate if it:

(a) shows that there is reasonable cause to believe that the dose received by the classified person was much greater than or much less than the dose recorded in the dose record; and

(b) includes sufficient information to permit a reliable reconstruction of the exposure conditions for the person during the relevant dose assessment period.

The investigation report should at least include the information in (a) and (b).

423 This regulation requires the employer to undertake an investigation if there is a reason to believe that the dose recorded in a classified person's dose record is substantially incorrect. For example, the employer may discover that a classified person has not been wearing a dosemeter at certain times when undertaking work with ionising radiation. Conversely, the employer may find that a batch of dosemeters was inadvertently exposed to ionising radiation during transit or storage.

424 The dose record entry might relate to an assessment by the ADS of whole-body or part-body dose, the neutron component of dose, or committed dose from intakes of radionuclides. Any change to a single entry will affect cumulative totals in recorded doses for the calendar year so far. HSE's consent is required in certain cases where the cumulative total exceeds a certain level (regulation 22(8)).

425 Further advice is given in an HSE information document on special entries[31] which has been made available to approved dosimetry services and can be downloaded from HSE's web site.

Deciding whether doses are much greater than or much less than the recorded dose

426 The ACOP guidance in paragraph 420 explains the circumstances where an investigation would be needed. There may be other circumstances where the employer considers that the dose received was much greater or much less than that recorded in the dose record, for example where a series of dose entries for consecutive dose assessment periods is affected. The employer will have to follow the procedure in regulation 22(3) in these cases, subject to the restrictions in regulations 22(5) and (8).

Unassessed components of dose

427 The provision for making alterations to recorded doses (special entries) only applies where dose assessments have already been recorded in the dose record. If the employer finds that a component of dose (eg dose to the hand) not previously assessed by an ADS is, in fact, significant, estimates of this new (and unrecorded) component may be added to the dose record by the ADS responsible for maintaining that record. The requirements of regulation 22(3) will not apply as none of the existing entries in the dose record relating to assessed doses needs to be changed.

(2)   In such a case, the employer shall -

(a)   take all reasonably practicable steps to inform each person for whom a dose assessment has been made of the result of that assessment; and

(b)   keep a record of the assessment or a copy thereof until the person to whom the record relates has or would have attained the age of 75 years but in any event for at least 50 years from the date of the relevant accident.

General

438   In most cases, people likely to be involved in incidents of the type covered by regulation 23 will either be:

(a)   classified persons who are subject to systematic dose assessments under regulation 21(2); or

(b)   employees who have been issued with a dosemeter or other device (eg criticality dosemeter) provided by an ADS under regulation 12(2).

439   In general, dosimetry services approved for the assessment of doses from external radiation are capable of providing a prompt assessment of dose, on request, in the event of exposure to ionising radiation arising from accidents or other occurrences. In special cases where life-threatening doses might be received, a few dosimetry services have been specially approved for assessing doses from external radiation which may exceed 0.5 gray. In these situations it is vital to provide early dose information to medical officers about those people who may have received an exposure which could be life-threatening.

440   It is important for the employer to advise the ADS responsible for assessing doses as early as practicable when that employer becomes aware that an exposure may have occurred as a result of an accident or other incident. The employer will then need to arrange for dosemeters, other devices, or bio-assay samples to be dispatched without undue delay to that ADS for analysis. If doses of the order of 0.5 gray or above are suspected, time will normally be of the essence. Where appropriate, the employer should arrange for the ADS to identify the dose as an accidental dose in the dose record (see regulation 25).

441   In cases where regulation 23(1)(c) applies, appropriate means for the assessment of dose may include:

(a)   examination of biological specimens, for example hair, nail clippings, blood etc; and

(b)   computation of dose from measured dose rates or contamination levels together with a knowledge of exposure times in the area and distance from the place of measurement, depending on the advice of the RPA.

442   The employer should normally seek advice from the RPA about the investigation (see ACOP advice in paragraph 217) and, where appropriate, the appointed doctor (or employment medical adviser) who undertakes medical surveillance in accordance with regulation 24(2) (see also ACOP advice in paragraph 448).

Application to medical exposures

443   This regulation does not apply to the protection of those undergoing a medical examination or treatment.

Medical surveillance

(1)   This regulation shall apply in relation to -

(a)   classified persons and persons whom an employer intends to designate as classified persons;

(b)   employees who have received an overexposure and are not classified persons;

(c)   employees who are engaged in work with ionising radiation subject to conditions imposed by an appointed doctor or employment medical adviser under paragraph (6).

When is medical surveillance required?

444   The employer of any individual exposed to ionising radiations as a result of work activities has the responsibility for deciding when such an employee needs to be designated as a classified person in accordance with regulation 20. Before that person is classified, the employer will need to ensure that the employee has been certified as fit for the intended type of work within the last 12 months. This may require a medical examination (see ACOP advice in paragraph 446). The employer will then need to make arrangements with the appointed doctor (or employment medical adviser) for continuing medical surveillance. Also, the employer will need to arrange for adequate medical surveillance for any employee who has received an overexposure, whether or not that employee has been designated as a classified person (see also guidance to regulation 25).

Appointed doctor and employment medical advisers

445   'Appointed doctor' and 'employment medical adviser' are defined in regulation 2(1). Employers may obtain the names of appointed doctors by contacting their local HSE office. Employment medical advisers (EMAs) are also known as medical inspectors.

(2)   The employer shall ensure that each of his employees to whom this regulation relates is under adequate medical surveillance by an appointed doctor or employment medical adviser for the purpose of determining the fitness of each employee for the work with ionising radiation which he is to carry out.

446   Adequate medical surveillance should include:

(a)   a medical examination before first being designated as a classified person in a post involving work with ionising radiations;

(b)   periodic reviews of health at least once every year;

(c)   special medical surveillance of an employee when a relevant dose limit has been exceeded;

(d)   determining whether specific conditions are necessary; and

(e)   a review of health after cessation of work where this is necessary to safeguard the health of the individual.

447   The nature of the medical surveillance for each individual should take account of the nature of the work with ionising radiation and that individual's state of health.

448 Medical surveillance carried out following an investigation under regulation 25 should include a special medical examination of the individual if that person has received an effective dose of ionising radiation in excess of 100 millisieverts in a year or an equivalent dose of at least twice any relevant annual dose limit.

Purpose of medical surveillance

449   The main purpose of medical surveillance is to determine an individual's fitness or continuing fitness for the intended work with ionising radiation. In this context, fitness of the person is not restricted to possible health effects from exposure to ionising radiation. The appointed doctor (or employment medical adviser) will need to take account of specific features of the work with ionising radiation, such as the fitness of the individual:

(a)   to wear any personal protective equipment (including respiratory protective equipment) required to restrict exposure;

(b)   with a skin disease, to undertake work involving unsealed radioactive materials; and

(c)   with serious psychological disorders, to undertake work with radiation sources that involve a special level of responsibility for safety.

450   In some posts, the nature of the work may be such that an employee is at risk from acute exposure to high levels of external radiation (eg site radiography) as a result of an accident. In such cases, the appointed doctor (or employment medical adviser) may take into account this potential for acute exposures to ionising radiation in deciding what level of medical surveillance is appropriate, even where recorded doses are generally low.

451   For individuals who have received an overexposure, medical surveillance (including monitoring for possible biological effects) is mainly intended to assess fitness to continue the work with ionising radiation.

Medical examination before designation as a classified person

452   In general, a medical examination will not be needed when a classified person changes employment, provided that the person has been certified fit for that type of work with ionising radiation within the preceding 12 months. A copy of that certification should be obtained from the previous employer and kept in the health record. The appointed doctor (or employment medical adviser) may also obtain previous clinical information, with the co-operation of the doctor who last undertook medical surveillance for the individual. Any conditions on the individual's work already imposed by the previous doctor would continue to have effect, at least until the next periodic review (see paragraphs 462-464).

453   Where there is a change of work which will involve exposure to a different risk from ionising radiations, the appointed doctor (or employment medical adviser) may decide that a medical examination is necessary in order to determine whether the person is fit for work in the new duties. The appointed doctor may consider that an examination is necessary, for example if the employee changes duties involving work with sealed sources to work with unsealed radionuclides, or has to use personal protective equipment for the first time.

Special medical examinations

454   The ACOP advice in paragraph 446 explains that special medical surveillance is usually necessary for any employee who had received an overexposure (and is subject to an investigation under regulation 25). The medical surveillance may include a special medical examination if the appointed doctor (or employment medical adviser) considers this to be necessary in the circumstances. The special medical examination may involve counselling the individual and detailing possible restrictions on further exposure. At the dose level specified in paragraph 448 special tests, such as chromosome aberration analysis, may be warranted to help establish the degree of any overexposure.

Periodic reviews of health

455   After the initial medical examination conducted prior to designation as a classified person (regulation 24(2) and ACOP advice in paragraph 446), periodic reviews of health should take place at least once every year. The appointed doctor (or employment medical adviser) may specify a shorter period between reviews.

456   The format of the review will be a matter of judgement on the part of the appointed doctor (or employment medical adviser), who will take into account any guidance issued to appointed doctors by HSE. The review will involve at least an assessment of the dose profile for the individual (as provided by the employer under regulation 24(8)) and sickness absence records (see paragraphs 466-468). The appointed doctor (or employment medical adviser) may also need access to other relevant records concerning the working conditions of the classified person, for example records of monitoring kept in accordance with regulation 19. Periodic reviews may also involve an interview with the individual and occasionally a medical examination and medical tests, depending on such factors as the nature of the work and the individual's state of health. The employer is responsible for making any necessary arrangements for medical surveillance. If the doctor wishes to see the individual as part of the periodic review, the employer should arrange for that person to see the doctor at an appropriate time during working hours (see regulation 34(5)).

Provision of facilities

457   The employer should either provide suitable facilities for the appointed doctor (or employment medical adviser) or allow employees to visit the doctor's surgery or examination room. The employer should bear the full cost, including time off for attendance.

Health record

(3)   The employer shall ensure that a health record, containing the particulars referred to in Schedule 7, in respect of each of his employees to whom this regulation relates is made and maintained and that that record or a copy thereof is kept until the person to whom the record relates has or would have attained the age of 75 years but in any event for at least 50 years from the date of the last entry made in it.

458   Employers may use any format for the health record provided that it contains at least the particulars in Schedule 7. Confidential clinical information should not be recorded in the health record but kept in a suitable record by the appointed doctor (or employment medical adviser). The Data Protection Act 1998[29] contains data protection requirements relevant to medical surveillance records. These requirements include the right of data subjects to see their health records.

459 When dose rates to the abdomen are not likely to exceed 13 millisieverts in any three-month interval and the employer has completed the relevant part of the health record to that effect before each review, the question of whether that employee is a woman of reproductive capacity does not arise for the purpose of medical surveillance. If appropriate, the appointed doctor (or employment medical adviser) will specify as a special condition for that employee that she is subject to the additional dose limit for a woman of reproductive capacity.

Valid entries in health record for periodic reviews of health

(4)   Subject to paragraph (5), the employer shall ensure that there is a valid entry in the health record of each of his employees to whom this regulation relates (other than employees who have received an overexposure and who are not classified persons) made by an appointed doctor or employment medical adviser and an entry in the health record shall be valid -

(a)   for 12 months from the date it was made or treated as made by virtue of paragraph (5);

(b)   for such shorter period as is specified in the entry by the appointed doctor or employment medical adviser; or

(c)   until cancelled by an appointed doctor or employment medical adviser by a further entry in the record.

(5)   For the purposes of paragraph (4)(a), a further entry in the health record of the same employee shall, where made not less than 11 months nor more than 13 months after the start of the current period of validity, be treated as if made at the end of that period.

460 Reviews of health should take place annually (unless the appointed doctor or employment medical adviser judges they are needed more frequently in particular cases). However, regulation 24(5) allows some leeway in the timing of reviews to avoid the need for entries to be made on the exact anniversary of previous ones. The review may be carried out up to one month before or one month after the due date but treated as if it had been carried out 12 months since the last review. For example, if a review which is due on 10 July is carried out between 10 June and 10 August, the next review will be due on 10 July the following year.

461 The doctor should make the signed entry in the record, not the employer (see Schedule 7). The contact details of the doctor who made the entry would normally be given in the health record.

Employees certified unfit /fit subject to conditions

(6)   Where the appointed doctor or employment medical adviser has certified in the health record of an employee to whom this regulation relates that in his professional opinion that employee should not be engaged in work with ionising radiation or that he should only be so engaged under conditions he has specified in the health record, the employer shall not permit that employee to be engaged in the work with ionising radiation except in accordance with the conditions, if any, so specified.

462 If appropriate, the appointed doctor (or employment medical adviser) will list any conditions in the health record which should be followed in order to ensure that the employee remains fit to continue the work with ionising radiation, including the type of work to which the conditions relate.

463   Any conditions specified in the health record must be followed until such time as they are rescinded by an appointed doctor (or employment medical adviser). If the doctor declares that a person is unfit for certain types of work with ionising radiation, the employer should not permit that person to undertake that work with ionising radiation.

464   The conditions specified in the health record may, for example, place restrictions on the type of work undertaken or the maximum dose of radiation a person should receive. The restrictions might be applied to the use of respiratory protective equipment or work with unsealed radioactive sources.

465   Regulation 24(9) allows employees to seek a review of any decision by the doctor if they are aggrieved by any condition recorded in the health record or a finding that the employee is not fit for the work with ionising radiation.

Information made available to the appointed doctor/employment medical adviser

(7)   Where, for the purpose of carrying out his functions under these Regulations, an appointed doctor or employment medical adviser requires to inspect any workplace, the employer shall permit him to do so.

(8)   The employer shall make available to the appointed doctor or employment medical adviser the summary of the dose record kept by the employer pursuant to regulation 21(7) and such other records kept for the purposes of these Regulations as the appointed doctor or employment medical adviser may reasonably require.

466   The records made available to the appointed doctor or employment medical adviser before the periodic review of health is undertaken should always include any relevant records of sickness absence for the person as well as the health record and copies of the summaries of the dose record provided by the approved dosimetry service and retained in accordance with regulation 21(7).

467   The appointed factory doctor must have direct access to dose summary information provided to the employer by the approved dosimetry service responsible for maintaining the dose record. The employer can make any mutually convenient arrangement for the appointed doctor (or employment medical adviser) to view this summary, which has to be retained for at least two years from the end of the year to which it relates; it need not be available as a paper record. The employer will want to make sure that the appointed doctor (or employment medical adviser) also has a copy of any explanation provided by the ADS of codes or abbreviations used in the dose summary. The health record should include the contact details of the ADS so the appointed doctor can resolve any difficulties with interpretation of the dose summary.

468   Normally, the appointed doctor (or employment medical adviser) would see copies of sickness absence records, automatically. The appointed doctor (or employment medical adviser) may also ask to see copies of other records kept for the purposes of these Regulations; it is reasonable for the employer to make these available to the appointed doctor (or employment medical adviser) provided sufficient notice is given. In addition, the appointed doctor (or employment medical adviser) may wish to review other information concerning the specific work with ionising radiation intended to be carried out, for example records of monitoring kept in accordance with regulation 19.

(9) *Where an employee is aggrieved by a decision recorded in the health record by an appointed doctor or employment medical adviser he may, by an application in writing to the Executive made within 3 months of the date on which he was notified of the decision, apply for that decision to be reviewed in accordance with a procedure approved for the purposes of this paragraph by the Health and Safety Commission, and the result of that review shall be notified to the employee and entered in his health record in accordance with the approved procedure.*

Regulation 25   Investigation and notification of overexposure

(1) *Where a radiation employer suspects or has been informed that any person is likely to have received an overexposure as a result of work carried out by that employer, that employer shall make an immediate investigation to determine whether there are circumstances which show beyond reasonable doubt that no overexposure could have occurred and, unless this is shown, the radiation employer shall -*

(a) *as soon as practicable notify the suspected overexposure to -*

    (i) *the Executive;*

    (ii) *in the case of an employee of some other employer, that other employer; and*

    (iii) *in the case of his own employee, the appointed doctor or employment medical adviser;*

(b) *as soon as practicable take reasonable steps to notify the suspected overexposure to the person affected; and*

(c) *make or arrange for such investigation of the circumstances of the exposure and an assessment of any relevant dose received as is necessary to determine, so far as is reasonably practicable, the measures, if any, required to be taken to prevent a recurrence of such overexposure and shall forthwith notify the results of that investigation and assessment to the persons and authorities mentioned in sub-paragraph (a) above and shall -*

    (i) *in the case of his employee, forthwith notify that employee of the results of the investigation and assessment, or*

    (ii) *in the case of a person who is not his employee, where the investigation has shown that that person has received an overexposure, take all reasonable steps to notify him of his overexposure.*

(2) *A radiation employer who makes any investigation pursuant to paragraph (1) shall make a report of that investigation and shall -*

(a) *in respect of an immediate investigation, keep that report or a copy thereof for at least 2 years from the date on which it was made; and*

(b) *in respect of an investigation made pursuant to sub-paragraph (c) of paragraph (1), keep that report or a copy thereof until the person to whom the record relates has or would have attained the age of 75 years but in any event for at least 50 years from the date on which it was made.*

(3)   Where the person who received the overexposure is an employee who has a dose record, his employer shall arrange for the assessment of the dose received to be entered into that dose record.

Application to medical exposures

469   This regulation does not apply to the protection of those undergoing a medical examination or treatment or to 'comforters and carers' because they are not subject to dose limits. However, it does apply to staff who carry out those exposures and members of the public.

Need for an investigation

470   The regulation applies to any overexposure, as defined, or suspected overexposure whether it arises from a single incident or because the total of doses received by the person during the period in question exceeds any dose limit for that period. It also includes situations where an individual made subject, under regulation 26, to a further restriction for the remainder of a dose limitation period, is suspected of receiving a dose in excess of that additional restriction.

Nature of immediate investigation

471   The main purpose of the immediate investigation is to rule out suspected incidents which it can readily be shown did not take place. For example, it might be suspected that an individual had been inadvertently exposed to the beam from an X-ray set, but the immediate investigation showed that the set was not in fact energised at the time. In cases where a significant exposure cannot be excluded, reviews intended to refine the initial estimate or assessment of the dose received are not part of the immediate investigation. Such reviews may be carried out, if appropriate, during the detailed investigation that follows.

Nature of the detailed investigation

472   The purpose of the detailed investigation under regulation 25(1)(c) is to establish why the overexposure occurred, what dose was received by the individual and what steps are necessary to prevent a recurrence of that overexposure. The radiation employer should normally consult the RPA about these investigations (see ACOP advice in paragraph 217), in addition to the person(s) affected. It may also be appropriate to seek the views of relevant safety representative(s) and any established safety committee.

473   The extent of the investigation will depend on the difficulty in establishing the cause of the overexposure and the magnitude of the dose(s) received. The Approved Dosimetry Service will need to record the assessed doses separately in the person's dose record as accidental doses, where appropriate.

474   If the individual who received the overexposure is an employee, the appointed doctor (or employment medical adviser) undertaking medical surveillance for that person should be involved at an appropriate stage in the investigation. The doctor may specify conditions for future work with ionising radiation (see paragraphs 446-448 and 462-464). If necessary, a health record should be opened for that employee (see paragraph 444). The appointed doctor (or employment medical adviser) may wish to seek the views of the employee concerned and take account of any likely future exposure of that person, as advised by the employer.

475 A typical detailed investigation is likely to include evidence about:

(a) the work routine of the individual, and immediate work colleagues, during the period concerned;

(b) details of any radiation monitors/personal alarms in use in the areas concerned during the period under investigation;

(c) involvement of the individual in any known incidents in which they may have received an unusual exposure;

(d) assessed or estimated doses over the last few years, compared to those of work colleagues undertaking similar work;

(e) results of any special radiation survey in the areas concerned (eg as part of a reconstruction advised by the RPA) to identify any deterioration in physical control measures;

(f) adherence to local rules or deficiencies in local rules;

(g) training, instruction or information received and general competence for the work undertaken; and

(h) other possible explanations for a suspected overexposure (eg there is evidence that the employee has continued to wear a dosemeter while receiving a medical exposure).

Report of investigation

476 The employer might wish to make the report or summary available to the established safety committee or to appointed safety representative(s) and at least a summary to the individual employees concerned.

Dose limitation for overexposed employees

(1)   Without prejudice to other requirements of these Regulations and in particular regulation 24(6), where an employee has been subjected to an overexposure paragraph (2) shall apply in relation to the employment of that employee on work with ionising radiation during the remainder of the dose limitation period commencing at the end of the personal dose assessment period in which he was subjected to the overexposure.

(2)   The employer shall ensure that an employee to whom this regulation relates does not, during the remainder of the dose limitation period, receive a dose of ionising radiation greater than that proportion of any dose limit which is equal to the proportion that the remaining part of the dose limitation period bears to the whole of that period.

(3)   The employer shall inform an employee who has been subjected to an overexposure of the dose limit which is applicable to that employee for the remainder of the relevant dose limitation period.

(4)   In this regulation, "dose limitation period" means, as appropriate, a calendar year or the period of five consecutive calendar years.

477 An employer may allow an employee who has received an overexposure to continue to work with ionising radiation, provided that:

(a) the provisions of regulation 25(1) have been fully complied with, by means of a detailed investigation etc;

(b)   the employee has been subject to medical surveillance for overexposed workers (see ACOP advice in paragraphs 446-448 and guidance in paragraph 454); and

(c)   the work is performed in accordance with any conditions imposed by the appointed doctor (or employment medical adviser) by an entry in the health record to ensure that the person remains fit for work with ionising radiation (see guidance in paragraphs 462-464).

478   Where an employee is subject to the five-year limit on effective dose, the employer will need to ensure that the employee does not receive more than the stated proportion of the five-year dose limit for each five-calendar-year period, which includes the calendar year in which the overexposure was received.

PART VI ARRANGEMENTS FOR THE CONTROL OF RADIOACTIVE SUBSTANCES, ARTICLES AND EQUIPMENT

Sealed sources and articles containing or embodying radioactive substances

(1)   Where a radioactive substance is used as a source of ionising radiation in work with ionising radiation, the radiation employer shall ensure that, whenever reasonably practicable, the substance is in the form of a sealed source.

479   A sealed source is one that under normal conditions of use prevents any dispersion of radioactive substances into the environment (see definition in regulation 2(1)). By requiring use of sealed sources, whenever reasonably practicable, the risks of dispersal are minimised.

(2)   The radiation employer shall ensure that the design, construction and maintenance of any article containing or embodying a radioactive substance, including its bonding, immediate container or other mechanical protection, is such as to prevent the leakage of any radioactive substance -

(a)   in the case of a sealed source, so far as is practicable; or

(b)   in the case of any other article, so far as is reasonably practicable.

480   The regulation applies not only to sealed sources but also to articles containing unsealed radioactive material. It requires the radiation employer to establish whether the design and construction of the source or article is fully fit for its intended use and to take account of the actual work to be done. Such considerations arise in any case from the assessment required by regulation 7.

481   Information from the manufacturers or suppliers of sources and articles about the mechanical protection used, including bonding materials, will allow the radiation employer to assess whether the manufacturing specification suits the intended use of the source. The overriding objective needs always to be prevention of loss of containment. For example, if the source is to be used in wet or aggressive conditions, the possibility of corrosion should be taken into account.

482   Where a sealed source reaches the end of the working life for the source capsule recommended by the supplier or manufacturer, a review of its condition is advised, with a view to replacing the source or having it examined by the supplier or manufacturer. If the source is not to be replaced, it is

advisable to set a time limit on its continued use after which a further review would normally be undertaken. Where the supplier or manufacturer does not specify a recommended working life, it may be advisable to carry out the first review within five years of manufacture of the source or to seek advice from the RPA about a period for review which is more appropriate in the circumstances.

Suitable leak tests

(3) Where appropriate, the radiation employer shall ensure that suitable tests are carried out at suitable intervals to detect leakage of radioactive substances from any article to which paragraph (2) applies and the employer shall make a suitable record of each such test and shall retain that record for at least 2 years after the article is disposed of or until a further record is made following a subsequent test to that article.

483 The purpose of a leak test is to show that the mechanisms for preventing dispersal of radioactive substances are functioning as intended. The assessment required by regulation 7 should identify potential ways in which containment could be lost and their likelihood of occurring. A test method and a frequency of testing should then be chosen that is capable of detecting leakage of radioactivity from the source or article before a radiation risk arises. Where testing is appropriate under normal operating conditions, the interval between tests should not exceed two years.

484 Tests for leakage should normally be carried out directly on the sealed source capsule or the article containing the radioactive substances, thus providing the best possible check for loss of containment. Where the source capsule or article is not accessible then an indirect test may be necessary, but conducted on parts that can reasonably be expected to have become contaminated in the event of leakage. Either way it is important for test methods to specify clear pass/fail criteria.

485 Another reason for choosing an indirect test is that a person conducting a direct test could receive an excessive radiation dose, in which case it can be done at the nearest routinely accessible position. However, any such decision needs to be balanced against the radiological implications of contamination resulting from the indirect method failing to detect a loss of containment. Also, particular care is needed in dismantling, or gaining closer than normal access to a source subject to indirect testing, because of the greater risk of contamination. This risk needs to be clearly flagged in the test record.

486 The manufacturer or supplier will normally advise about periodic leak testing and the methods to adopt to give the required level of assurance that radioactive material will not be allowed to disperse. In the absence of such advice, test methods set out in ISO 9978[32] may be appropriate.

487 Tests are not usually considered appropriate in the following circumstances:

(a)   where the sealed source contains solely gaseous radioactive substances;

(b)   on an article containing a radioactive substance which is solely designed and used for the purpose of detecting smoke or fire and is installed in a building for that purpose;

(c) where an article containing a radioactive substance, not being a sealed source, is by design open, for example a syringe, bottle or similar equipment; and

(d) on any sealed sources during irradiation in a nuclear reactor.

488 Leak tests may not be appropriate in other situations, eg on sources or articles which have no dimensions greater than 5 mm, for example gold grains and microspheres. These may be treated as dispersible radioactive substances. However, this advice is not intended to preclude leak tests being carried out on smaller sources where it is appropriate. Even where leak tests are not carried out, care is always needed to prevent contamination.

Suitable interval for leak tests

489 The ACOP guidance in paragraph 483 advises that the usual interval between tests should not normally exceed two years. It is recognised that there may occasionally be practical difficulties in ensuring that such tests are always carried out every two years, although this should normally be the aim. More frequent testing is recommended for situations where the radiological implications of a loss of containment could be severe, or the physical or chemical conditions are such that deterioration of the source or its containment might occur, for example in a hot and humid environment. Additional tests for leakage will also be appropriate if any damage is suspected, or where any work has been carried out which could have affected the integrity of the capsule or article.

490 Where a source or article is inaccessible in normal operation, leak testing can sometimes await occasions of improved access, such as scheduled maintenance, rather than proceeding with the routine test. This occurs with gauges installed inside process vessels or in bunkers in the mining industry, where gaining access during normal operation can pose a significant physical risk. Thus, the period between leak tests on a particular source does not always need to be the same and can be varied to allow for an improved test.

491 Increasing the frequency of leak testing is advisable when a sealed source is going to be retained in use beyond the recommended working life given to the source capsule by the supplier or manufacturer. Where there is no recommended working life, then the frequency of leak testing needs to be considered as part of the periodic reviews of its condition (see paragraph 482).

Suitable record of leak tests

492 The recommended contents of a suitable leak test record are:

(a) the identification of the source or article which is the subject of the test;

(b) the date of test;

(c) the reason for test (eg pre-use, manufacturer's test, nominal routine, after incident);

(d) the methods of test, including, when the source or article has not been tested directly, a statement of what part of the device was tested and a statement about whether this is likely to detect any leaking material. The description of the test will normally include a statement of the pass/fail criteria;

(e)   numerical results of the test;

(f)   the result of the test (pass or fail);

(g)   any action taken if the source failed the test; and

(h)   the name and signature of the person carrying out the test.

Accounting for radioactive substances

For the purpose of controlling radioactive substances which are involved in work with ionising radiation which he undertakes, every radiation employer shall take such steps as are appropriate to account for and keep records of the quantity and location of those substances and shall keep those records or a copy thereof for at least 2 years from the date on which they were made and, in addition, for at least 2 years from the date of disposal of that radioactive substance.

493   The procedures for accounting should ensure that the location of radioactive substances is known and, as a consequence, losses of significant quantities can quickly be identified. A frequency for checking the location of the source should be determined, taking account of the likely movement of the source, its potential for being displaced and its susceptibility to damage. For portable sources, such as radiography sources and portable gauges, the check should be at least on each working day.

494   Other examples of intervals at which the location of a source should be updated are:

(a)   for static sources securely attached to machines the interval between checks may be up to one month, providing that additional checks are carried out following any maintenance or repair which could have affected the source; and

(b)   for sources located within patients, the interval between checks should be compatible with the clinical treatment of that patient.

495   Employers registered under RSA93,[6] or operating under an RSA93 exemption order, may also be legally required to keep records. In these circumstances a single system of accounting can readily be adopted which satisfies both sets of legal requirements.

Accounting - general advice

496   The records for accounting for any particular radioactive substance will normally need to include:

(a)   a means of identification, which for sealed sources should usually be unique;

(b)   the date of receipt;

(c)   the activity at a specified date;

(d)   the whereabouts of the substance, updated at appropriate intervals; and

(e)   the date and manner of disposal (when appropriate).

116

497 An annual check is advisable to ensure that the accounting record is a true record. Such a check necessarily excludes any radioactive substances which have decayed to an insignificant quantity.

Accounting - specific situations

498 In production processes directly involving dispersible radioactive substances, the accounting procedures will vary considerably with the scale of operation. In most cases the records held for production processing and waste disposal should be sufficient.

499 In small-scale laboratories it should be sufficient to know the activity present and the radionuclides involved in each room, supported by the records required for waste disposal purposes.

500 Very small sources or articles having no dimension greater than 5 mm may be treated as dispersible radioactive substances for the purposes of accounting.

501 Accounting procedures may not be appropriate for radioactive substances where they:

(a)　have a half-life of less than three hours;

(b)　are in the form of contamination;

(c)　are in the form of a discrete source or confined radioactive substance, and their quantity or concentration does not exceed that specified in Schedule 8, columns 2 and 3;

(d)　are dispersed in the body of a person, though this dispensation does not apply to samples taken from such persons;

(e)　form part of a nuclear reactor, until such time as radioactive components are removed or the reactor is decommissioned; or

(f)　are undergoing irradiation in a nuclear reactor, particle accelerator or other similar device, provided they are not directly related to the operation of the plant.

Keeping and moving of radioactive substances

(1)　Every radiation employer shall ensure, so far as is reasonably practicable, that any radioactive substance under his control which is not for the time being in use or being moved, transported or disposed of -

(a)　is kept in a suitable receptacle; and

(b)　is kept in a suitable store.

502 In addition to this provision, employers registered under RSA93,[6] or exempted from registration by an exemption order, are legally required to keep radioactive substances properly. In these circumstances, a single system of keeping can readily be adopted which satisfies both sets of legal requirements.

Suitable receptacle

503 For radioactive substances not in use, a receptacle is suitable where it ensures effective restriction of exposure, prevention of dispersal and physical security. General characteristics of the receptacle which need consideration are:

(a) radiation shielding - it is advisable that the surface dose rate never exceeds 2 millisieverts per hour and usually it should be much less;

(b) its ability to withstand damage from normal use and foreseeable misuse;

(c) its fire resistance; and

(d) its design in relation to prevention of unauthorised exposure or dispersal.

504 Special considerations of receptacle design arise where:

(a) the radioactive substance is corrosive, self-heating or pyrophoric;

(b) there could be pressure build-up inside the receptacle; or

(c) the storage environment itself is corrosive.

Suitable store

505 Characteristics of a suitable store for radioactive substances are:

(a) protection from the effects of the weather;

(b) resistance to fire sufficient to minimise dispersal and loss of shielding, taking into account the combustibility of surrounding materials and the likely temperatures that would be reached in the event of a fire;

(c) shielding to achieve outside the store the lowest dose rate that is reasonably practicable. Where non-classified persons may approach the outside of the store it is advisable that the dose rate does not normally exceed 2.5 microsieverts per hour and may need to be lower in special cases if the radiation employer wishes to avoid designating the area as a supervised area (see regulation 16(3));

(d) ventilation to prevent significant accumulations of gases and vapours (whether radioactive or not) or of any accidentally dispersed radioactive substance; and

(e) proper physical security such that access is only normally possible to those people permitted by the employer, whether or not the inside of the store is a controlled area.

506 It is recommended that any store allocated to radioactive substances is reserved for such substances, their immediate containers and receptacles and ancillary items such as handling tools and shielding material. In particular, it is advisable that nothing explosive or highly flammable should be kept in the store.

507 A sign prominently displayed on the outside of the store (preferably on the door) serves to warn people in the vicinity that the store may contain radioactive substances. Such signs should conform to the Health and Safety (Safety Signs and Signals) Regulations 1996.[13]

118

Suitable receptacles for moving radioactive substances

(2)   Every employer who causes or permits a radioactive substance to be moved (otherwise than by transporting it) shall ensure that, so far as is reasonably practicable, the substance is kept in a suitable receptacle, suitably labelled, while it is being moved.

508   This requirement relates to suitable packaging and labelling of radioactive substances during movement other than transport. It therefore particularly applies during site movements of radioactive material.

509   Transport is defined in regulation 2 and essentially covers all conveyance through public places. Standards for packaging and labelling during transport are to be found in the Radioactive Material (Road Transport) (Great Britain) Regulations 1996[33] and the Packaging, Labelling and Carriage of Radioactive Material by Rail Regulations 1996.[34]

510   Considerations for judging suitability of a receptacle for movement are:

(a)   the adequacy of the shielding to protect the person moving the substances;

(b)   the distance of the movement;

(c)   the likely hazards to be encountered and the consequences of an incident, for example from spillage or dispersal; and

(d)   the physical and chemical form and activity of the substances being moved.

511   Labelling needs to provide sufficient information for the safety of the person moving the receptacle, and indicate the nature and activity of the substances being moved or allow their ready identification. The extent of the information will depend on circumstances, but taking appropriate action in the event of accidental spillage or dispersal would be one.

(3)   Nothing in paragraphs (1) and (2) shall apply in relation to a radioactive substance while it is in or on the live body or corpse of a human being.

Notification of certain occurrences

(1)   Every radiation employer shall forthwith notify the Executive in any case where a quantity of a radioactive substance which was under his control and which exceeds the quantity specified for that substance in column 4 of Schedule 8 -

(a)   has been released or is likely to have been released into the atmosphere as a gas, aerosol or dust; or

(b)   has been spilled or otherwise released in such a manner as to give rise to significant contamination.

(2)   Paragraph (1) shall not apply where such release -

(a)   was in accordance with a registration under section 10 of the Radioactive Substances Act 1993[(a)] or which was exempt from such registration by virtue of section 11 of that Act; or

(b) was in a manner specified in an authorisation to dispose of radioactive waste under section 13 of the said Act or which was exempt from such authorisation by virtue of section 15 of that Act.

(a) 1993 c.12.

Notifying accidental releases and spillage

512   The term 'release' includes accidental spillages of radioactive substances, such as bench spills in a laboratory, and the term 'atmosphere' covers the internal environment of buildings as well as the external atmosphere. In general, where there is release of a dusty solid, it can be assumed for the purposes of regulation 30(1)(a) that the amount of dust released into the atmosphere is one-thousandth of the total amount involved in the occurrence. However, this assumption does not hold where an exceptionally fine dust is involved or there are other reasons to judge it inappropriate, for instance the release has occurred under pressure or resulting from an explosion.

513   As a general rule, contamination arising from a spillage which exceeds the quantity specified in Schedule 8, column 4 would be regarded as significant. The exception to this is where the spillage is in an enclosure or other such localised facility, so designed, maintained and used as to effectively prevent the release going beyond that facility. This exception would apply, for instance, to glove boxes, purpose-designed enclosures and benches in laboratories and specially-designed toilets in nuclear medicine departments. However, this exception is not intended to cover releases affecting whole rooms or buildings where people work and could receive a significant exposure to ionising radiation as a result of the spillage.

Notifying other accidents etc

514   Regulation 32(6) also requires the relevant employer to notify HSE about certain incidents involving equipment used in connection with medical exposures. In addition the Reporting of Injuries, Diseases and Dangerous Occurrences Regulations 1995[35] require certain events to be reported to the relevant enforcing authority (usually HSE), in particular:

(a)   the malfunction of a 'radiation generator' or its ancilliary equipment used in fixed or mobile industrial radiography, the irradiation of food or the processing of products by irradiation, causes it to fail to de-energise at the end of the intended exposure period; or

(b)   the malfunction of equipment used in fixed or mobile industrial radiography or gamma irradiation causes a radioactive source to fail to return to its safe position by normal means at the end of the intended exposure period.

(3)   Where a radiation employer has reasonable cause to believe that a quantity of a radioactive substance which exceeds the quantity for that substance specified in column 5 of Schedule 8 and which was under his control is lost or has been stolen, the employer shall forthwith notify the Executive of that loss or theft, as the case may be.

515   Employers with substances registered under RSA93,[6] to keep radioactive material, or operating under an RSA exemption order, should also note that reporting a loss or theft to the relevant environment agency is a legal requirement.

516 Where accounting estimates involve large quantities of radioactive substances, as for example in the nuclear industry, uncertainties in those estimates can be of the order of the quantities specified in Schedule 8, column 5, thus apparently requiring notification to HSE under the terms of regulation 30(3). In such cases an agreement may be reached with HSE at local level as to what accountancy error is sufficiently large to constitute reasonable grounds for believing that the radioactive substances have been lost.

(4) Where a radiation employer suspects or has been informed that an occurrence notifiable under paragraph (1) or (3) may have occurred, he shall make an immediate investigation and, unless that investigation shows that no such occurrence has occurred, he shall forthwith make a notification in accordance with the relevant paragraph.

(5) A radiation employer who makes any investigation in accordance with paragraph (4) shall make a report of that investigation and shall, unless the investigation showed that no such occurrence occurred, keep that report or a copy thereof for at least 50 years from the date on which it was made or, in any other case, for at least 2 years from the date on which it was made.

Duties of manufacturers etc of articles for use in work with ionising radiation

(1) In the case of articles for use at work, where that work is work with ionising radiation, section 6(1) of the Health and Safety at Work etc. Act 1974(a) (which imposes general duties on manufacturers etc. as regards articles and substances for use at work) shall be modified so that any duty imposed on any person by that subsection shall include a duty to ensure that any such article is so designed and constructed as to restrict so far as is reasonably practicable the extent to which employees and other persons are or are likely to be exposed to ionising radiation.

(a) 1974 c.37; section 6 was amended by the Consumer Protection Act 1987 (c.43), section 36 and Schedule 3.

Application to medical exposures

517 Regulation 31(1) does not apply to the protection of those undergoing a medical examination or treatment. However, it does apply to the exposure of staff who carry out those exposures, members of the public, and to 'comforters and carers' as defined in regulation 2(1).

Duties of manufacturer and supplier to restrict exposure

518 The manufacturer, supplier or importer of any article embodying or containing a radioactive substance, including a sealed source, should ensure that suitable leak tests are carried out as soon as practicable after manufacture or importation.

519 The extension of the duty on manufacturers and suppliers under section 6(1) of the Health and Safety at Work etc Act 1974[2] with regard to articles for use at work is not intended to apply to equipment used for medical exposures. This equipment will be subject to the requirements of the Medical Devices Regulations 1994,[36] which are generally enforced by the Medical Devices Agency.

520 All people involved in the chain from the design, manufacture and supply to the use or installation of an article have a responsibility to ensure that unambiguous and comprehensive information is passed along that chain. Only then can the user know precisely how the article should be used so that standards for restricting exposures (and achieving compliance with these Regulations) can be maintained or improved commensurate with the article performing the function for which it was intended.

521 Sometimes a commercially available article that was neither designed for use in connection with work with ionising radiation, nor sold by the supplier for that purpose, is nevertheless used as such by a radiation employer. One example would be an extractor fan bought merely as a fan of known capacity from a retailer and then incorporated into a ventilation system intended to control airborne radioactive substances. Unless an article has been supplied on the basis of its future use or its design criteria, the radiation employer will want to ensure that it complies with the Regulations and achieves the necessary performance standard.

Critical examination by installer/erector

(2)   Where a person erects or installs an article for use at work, being work with ionising radiation, he shall -

(a)   where appropriate, undertake a critical examination of the way in which the article was erected or installed for the purpose of ensuring, in particular, that -

(i)   the safety features and warning devices operate correctly; and

(ii)   there is sufficient protection for persons from exposure to ionising radiation;

(b)   consult with the radiation protection adviser appointed by himself or by the radiation employer with regard to the nature and extent of any critical examination and the results of that examination; and

(c)   provide the radiation employer with adequate information about proper use, testing and maintenance of the article.

522 It is appropriate to carry out a critical examination if there may be radiation protection implications arising from the way in which an article is being or has been erected or installed.

523 Matters on which the radiation protection adviser should be consulted include the plans for installing the equipment, the nature and extent of any tests undertaken as part of the critical examination and the acceptability of any test results.

524 The critical examination may be carried out following erection or installation, during commissioning, or as part of trials prior to normal use, and may require co-operation between the various employers involved at each stage of a complex installation (see guidance on regulation 15). The duty to ensure that the critical examination is carried out rests with the employer who erects or installs the article, not the user. This remains the case even if the critical examination is carried out, by agreement, during final trials under the supervision of the user's own RPA. Unlike regulation 31(1), the requirement

to undertake a critical examination requires the duty holder to consider the protection provided for people undergoing medical exposures, as well as the adequacy of protection for staff and members of the public.

525   Although the erector or installer must consult an RPA, regulation 31(2) does not require that adviser to be present when the critical examination is carried out if this is unnecessary for the purpose of conducting the examination. The RPA may be the installer's own RPA or the RPA appointed by the employer who has purchased the equipment.

526   A critical examination should be carried out for articles containing radioactive substances and for X-ray generators. Other articles which may form part of plant should also be covered by the critical examination. This should take into account matters such as shielding, ease of decontamination of surfaces, containment and any other aspects of radiation protection. However, the critical examination need not include matters related to the intrinsic safety of the article which are not likely to be affected by the way in which the article has been erected or installed.

31(2)

Regulation 32

Equipment used for medical exposure

Regulation

(1)   Every employer who has to any extent control of any equipment or apparatus which is used in connection with a medical exposure shall, having regard to the extent of his control over the equipment, ensure that such equipment is of such design or construction and is so installed and maintained as to be capable of restricting so far as is reasonably practicable the exposure to ionising radiation of any person who is undergoing a medical exposure to the extent that this is compatible with the intended clinical purpose or research objective.

32(1)

Guidance

General advice on equipment used for medical exposure

527   This provision covers all equipment used in connection with medical exposures where the design, construction, installation, maintenance, and any fault that might develop in it, can affect the magnitude or the distribution of the absorbed dose received by the person undergoing a medical exposure. It includes equipment intended to be used in connection with diagnostic or therapeutic procedures using ionising radiation and interventional radiology. It covers both radiation equipment, as defined in regulation 32(8), and ancillary equipment such as image intensifiers, intensifying screens, film cassettes, digital fluorography systems, couches, anti-scatter grids, beam modifiers (eg filters and wedges), dose calibrators and film processing units.

528   Control measures to restrict the exposure of staff and members of the public from exposure to radiation emitted by the equipment (eg by the provision of adequate shielding and use of collimation) are covered elsewhere in these Regulations (see regulations 8-10 and 31).

Design and construction

529   The manufacturer of the equipment has duties under the Medical Devices Regulations 1994[36] (which implement Council Directive 93/42/EEC[37]) concerning the design and manufacture of equipment intended to be used for medical exposures. Regulation 32(1) of IRR99[1] requires the employer to consider restriction of patient exposure when purchasing equipment. The employer will also need to review, periodically, the capability

32(1)

of existing equipment to restrict exposure so far as reasonably practicable, consistent with its intended use. The standard of design and construction for some older existing equipment may not permit exposure to be restricted to a sufficient extent for some types of medical exposure.

530   Medical equipment bearing the CE marking when placed on the market is declared by the manufacturer to be in compliance with the essential requirements of the Medical Devices Regulations 1994.[37] Consequently, such equipment can be taken as satisfying the present requirement relating to design and construction, unless there are reasonable grounds for suspecting that the device does not comply, provided that the equipment is appropriate for the intended purpose.

531   Decisions on the purchase of ancillary equipment can have an important influence on the extent of medical exposures. For example, in diagnostic radiology the choice of a sensitive film/screen combination and low attenuation materials for ancillary equipment such as anti-scatter grids, film cassettes and table tops can have a significant impact on the magnitude of medical exposures.

Installation of equipment

532   The employer may find the report of the critical examination undertaken by the installer of the equipment useful in helping to ensure that the equipment has been properly installed for the intended clinical or research purpose (see regulation 31(2)). However, the employer should bear in mind that the critical examination is limited in scope to possible radiation protection implications arising from the way in which it was erected or installed. The critical examination is not intended to be a test of the initial integrity of the equipment.

Selection of equipment

533   The employer in control of equipment used for medical exposure needs to ensure that the equipment available for the range of radiological examinations or treatments to be undertaken is suitable for the purposes of complying with regulation 32(1). Arrangements should be made to ensure that users are able to select the most appropriate equipment for each medical exposure, and that equipment is not used for procedures for which it is not suitable (eg a dental X-ray set would not be appropriate for taking a radiograph of the chest).

Maintenance of equipment

534   'Maintained' is defined under regulation 2(1) in relation to equipment. The person who erects or installs the equipment should provide adequate information about any necessary maintenance (see regulation 31) and, under the Medical Devices Regulations 1994[37] the manufacturer must also provide information about maintenance.

535   Mechanical and electrical safety checks are an important part of maintenance. The requirements of the Provision and Use of Work Equipment Regulations 1998[38] (and associated ACOP) and the Electricity at Work Regulations 1989[39] will be relevant.

Indication of quantity of radiation produced for new equipment

(2)   An employer who has to any extent control of any radiation equipment which is used for the purpose of diagnosis and which is installed after the date of the coming into force of these Regulations shall, having regard to the extent of his control over the equipment, ensure that such equipment is provided, where practicable, with suitable means for informing the user of that equipment of the quantity of radiation produced by that equipment during a radiological procedure.

536   Suitable means may include, for example, a device which shows the product of X-ray tube current and exposure time, or a specially-designed ionisation chamber which indicates the dose-area product. Such devices may not be practicable for certain types of equipment, such as dental X-ray sets and equipment used in nuclear medicine.

537   This provision applies to radiation equipment as defined in regulation 32(8) and only to equipment installed for the first time after 1 January 2000.

Provision of a quality assurance programme for equipment

(3)   Every employer in respect of whom a duty is imposed by paragraph (1) shall, to the extent that it is reasonable for him to do so having regard to the extent of his control over the equipment, make arrangements for a suitable quality assurance programme to be provided in respect of the equipment or apparatus for the purpose of ensuring that it remains capable of restricting so far as is reasonably practicable exposure to the extent that this is compatible with the intended clinical purpose or research objective.

(4)   Without prejudice to the generality of paragraph (3), the quality assurance programme required by that paragraph shall require the carrying out of -

(a)   in respect of equipment or apparatus first used after the coming into force of this regulation, adequate testing of that equipment or apparatus before it is first used for clinical purposes;

(b)   adequate testing of the performance of the equipment or apparatus at appropriate intervals and after any major maintenance procedure to that equipment or apparatus;

(c)   where appropriate, such measurements at suitable intervals as are necessary to enable the assessment of representative doses from any radiation equipment to persons undergoing medical exposures.

538   A suitable quality assurance programme is one that establishes those planned and systematic actions necessary to provide adequate confidence that equipment will satisfy the requirements of regulation 32(1). The extent of the programme will depend on the nature and range of equipment in use. In drawing up a quality assurance programme, the employer should make it clear who has responsibility for organising the various elements, carrying out testing or dose assessment and for acting on any adverse findings.

539   The programme should specify the frequency of any testing (and other measurements) and appropriate action levels for equipment or

apparatus which is subject to periodic testing. If these levels are found to have been exceeded the employer should assess what remedial action is needed, including removal from service where necessary, taking into account the risk arising from its continued use for specified purposes. In establishing these levels, the employer should take into account guidance established by relevant professional bodies about criteria of acceptability for such equipment.

540 In devising a suitable quality assurance programme for equipment, the employer should give special attention to equipment used for medical exposure:

(a) of children;

(b) as part of a health screening programme;

(c) involving high doses to the patient such as interventional radiology, computed tomography or radiotherapy.

541 The employer should normally consult the appointed RPA about a suitable quality assurance (QA) programme for equipment used in connection with medical exposures (see ACOP advice in paragraph 217).

542 The purposes of testing before first use are:

(a) to ensure that the equipment is in a satisfactory condition to be used for medical exposures; and

(b) to provide baseline values against which the results of subsequent routine performance tests can be compared.

543 Periodic testing is also required to enable any deterioration in performance to be detected and if necessary corrected. Testing is also required after any major maintenance procedure, whether or not that procedure involved parts of the equipment which are directly relevant to radiation protection of the patient. It is necessary to undertake a test in these circumstances because such maintenance work may have required safety devices to be disabled temporarily or it may have affected the radiation protection of patients indirectly.

544 Testing may be carried out by either the employer's own staff or by contractors, or both. This will depend on the degree of expertise and the availability of equipment needed to conduct the tests. In any case, it is the employer's responsibility to ensure that the tests are adequate and that appropriate remedial action is taken when necessary. The employer will need to decide what measurements are to be made, the accuracy required and what levels should be set to identify any necessary remedial action.

545 The employer should generally identify appropriate test equipment as part of the QA programme. The employer will need to make arrangements to ensure that this is provided and maintained and that it is calibrated to an appropriate standard prior to use and at suitable intervals.

546 The results of any initial and periodic checks or tests will need to be properly documented as part of the QA programme.

Radiodiagnostic equipment and interventional radiology

547 It will be appropriate to undertake periodic measurements for the purpose of assessing representative patient doses where X-ray equipment is used for diagnosis or interventional radiology. However, this information would be of limited use for specialised equipment such as radiotherapy simulators. In deciding on an appropriate frequency for the tests, the employer should take account of relevant recommendations from professional bodies. Measurements should generally be made on a sample basis relating to individuals who are representative of the relevant population (ie of average size and weight). The QA programme should include adequate calibration of any devices or dosemeters provided to assess these doses.

Radiotherapy equipment

548 For equipment used in connection with radiotherapy, including treatment simulators, the aims of medical exposure are rather different and the QA programme should clearly be designed to support those aims. Important features of the programme are likely to include:

(a) initial and periodic calibration of radiation equipment and dosemeters; and

(b) constancy checks of the geometrical aspects of external therapy machines and simulators, beam uniformity, timers, and source positioning devices.

549 There is no need to assess patient doses for equipment intended for therapeutic use to comply with regulation 32(4). Dose assessments for each person undergoing radiotherapy will be covered by implementation of the individual treatment plan, which is outside the scope of this regulation.

Prevention of equipment failures

(5) Every employer who has to any extent control of any radiation equipment shall take all such steps as are reasonably practicable to prevent the failure of any such equipment where such failure could result in an exposure to ionising radiation greater than that intended and to limit the consequences of any such failure.

550 Employers will have considered the possibility of equipment faults occurring when purchasing new equipment. Also, preventive maintenance under regulation 32(1) should help to prevent failures due to routine wear and tear. To comply with this requirement the employer will need to consider such matters as the provision of safety devices (eg interlocks) to prevent incidents occurring as a result of a defect in or malfunction of equipment, especially radiotherapy equipment.

551 The employer may become aware of defects in equipment from information provided by the supplier or by other users or as a result of in-house operational experience. In these cases, the employer will need to assess whether any further action is necessary (in consultation with the supplier and the RPA) in the light of the national arrangements for reporting adverse incidents (see paragraph 557).

552 If the failure of a single component can give rise to unintended exposure of the patient or research subject the employer may need to take additional steps. These may involve ensuring that the exposure is automatically terminated within an appropriate preset time, tube current-time product (milliamp-seconds), or dose, to protect the individual undergoing the exposure. Where this is not reasonably practicable, it may be necessary to

ensure that such a failure is immediately detectable so action can be taken by the operator. The employer might also need to provide a contingency plan for responding to equipment failure or malfunction in particular cases, both to protect members of staff (as required by regulation 12) and people undergoing medical exposures.

553 Equipment may malfunction or not perform as intended as a result of misunderstandings between service engineers and users. Reasonable steps to prevent unintended exposures may therefore include clear hand-over procedures for equipment which has undergone maintenance.

Investigation of incidents

(6)   Where a radiation employer suspects or has been informed that an incident may have occurred in which a person while undergoing a medical exposure was, as the result of a malfunction of, or defect in, radiation equipment under the control of that employer, exposed to ionising radiation to an extent much greater than that intended, he shall make an immediate investigation of the suspected incident and, unless that investigation shows beyond reasonable doubt that no such incident has occurred, shall forthwith notify the Executive thereof and make or arrange for a detailed investigation of the circumstances of the exposure and an assessment of the dose received.

(7)   A radiation employer who makes any investigation in accordance with paragraph (6) shall make a report of that investigation and shall -

(a)   in respect of an immediate report, keep that report or a copy thereof for a period of at least 2 years from the date on which it was made; and

(b)   in respect of a detailed report, keep that report or a copy thereof for a period of at least 50 years from the date on which it was made.

(8)   In this regulation 'radiation equipment' means equipment which delivers ionising radiation to the person undergoing a medical exposure and equipment which directly controls the extent of the exposure.

554 HSE has published specific guidelines[40] on doses which are likely to be much greater than intended for particular types of medical exposure. If it is suspected that an incident of the type specified in regulation 32(6) has occurred, the employer in control of the equipment should arrange for an immediate investigation and notify HSE after this initial investigation, unless this shows that an equipment defect or malfunction did not occur.

555 The subsequent detailed investigation should normally be aimed at:

(a)   establishing what happened;

(b)   identifying the defect or malfunction and its causes;

(c)   deciding on remedial action to prevent a recurrence (regulation 32(5)); and

(d)   estimating the doses received by those patients or research subjects involved in the incident.

The ACOP guidance in paragraph 217 advises that the employer should normally consult the RPA about such investigations.

556 Regulation 32 is aimed at the prevention of exposures that are greater than intended. Exposures that are lower than intended can have serious consequences, especially for therapy procedures. However, the Regulations are only concerned with preventing unnecessary risks to people as a direct consequence of exposure to ionising radiation, not with the successful or unsuccessful treatment of medical conditions. Therefore, while action may be required under regulation 32(5) to minimise defects or malfunctions in equipment, an investigation under this regulation is not formally required if the malfunction or defect results in an exposure less than intended.

557 The Medical Devices Regulations 1994[37] require manufacturers to report to the Medical Devices Agency any incident which led to, or might have led to, death or serious injury. Therefore, it is vital that users report any such incident to manufacturers so they can fulfil their legal obligations. Also, the Medical Devices Agency operates a scheme for users to report any adverse incident concerning medical devices. Prompt reporting enables the Medical Devices Agency to take appropriate action with the manufacturer.

Misuse of or interference with sources of ionising radiation

No person shall intentionally or recklessly misuse or without reasonable excuse interfere with any radioactive substance or any electrical equipment to which these Regulations apply.

PART VII DUTIES OF EMPLOYEES AND MISCELLANEOUS

Duties of employees

(1) An employee who is engaged in work with ionising radiation shall not knowingly expose himself or any other person to ionising radiation to an extent greater than is reasonably necessary for the purposes of his work, and shall exercise reasonable care while carrying out such work.

(2) Every employee who is engaged in work with ionising radiation and for whom personal protective equipment is provided pursuant to regulation 8(2)(c) shall -

(a) make full and proper use of any such personal protective equipment;

(b) forthwith report to his employer any defect he discovers in any such personal protective equipment; and

(c) take all reasonable steps to ensure that any such personal protective equipment is returned after use to the accommodation provided for it.

(3) It shall be the duty of every outside worker not to misuse the radiation passbook issued to him or falsify or attempt to falsify any of the information contained in it.

(4) Any employee to whom regulation 21(1) or regulation 12(2)(b) relates shall comply with any reasonable requirement imposed on him by his employer for the purposes of making the measurements and assessments required under regulation 21(1) and regulation 23(1).

(5)   An employee who is subject to medical surveillance under regulation 24 shall, when required by his employer and at the cost of the employer, present himself during his working hours for such medical examination and tests as may be required for the purposes of paragraph (2) of that regulation and shall provide the appointed doctor or employment medical adviser with such information concerning his health as the appointed doctor or employment medical adviser may reasonably require.

(6)   Where an employee has reasonable cause to believe that -

(a)   he or some other person has received an overexposure;

(b)   an occurrence mentioned in paragraph (1) or (3) of regulation 30 has occurred; or

(c)   an incident mentioned in regulation 32(6) has occurred,

he shall forthwith notify his employer of that belief.

558   This regulation places duties on employees when they are engaged in some way in work with ionising radiation. The first paragraph requires the exercise of reasonable care generally. The employer has responsibilities under regulation 14 to ensure that employees are made aware of these legal duties.

Application to medical exposures

559   This regulation does not apply to the protection of those undergoing a medical examination or treatment. However, it does apply to staff who carry out those exposures.

Other duties of employees

560   Subsequent paragraphs relate to specific procedures and provisions of the Regulations, as follows:

(a)   use of personal protective equipment;

(b)   care of and not falsifying the radiation passbook by outside workers;

(c)   co-operation with the employer over dose measurements and assessments;

(d)   co-operation with the employer and doctor in medical surveillance; and

(e)   informing the employer about actual or suspected incidents which the employer has a duty to investigate, such as apparent overexposures or loss of a source.

The connection between these varied situations is that the employer can only meet certain specific duties under the Regulations with the co-operation of employees.

Approval of dosimetry services

(1)   The Executive (or such other person as may from time to time be specified in writing by the Executive) may, by a certificate in writing, approve (in accordance with such criteria as may from time to time be specified by the Executive) a suitable dosimetry service for such of the purposes of these Regulations as are specified in the certificate.

(2)   A certificate made pursuant to paragraph (1) may be made subject to conditions and may be revoked in writing at any time.

(3)   The Executive (or such other person as may from time to time be specified in writing by the Executive) may at such suitable periods as it considers appropriate carry out a re-assessment of any approval granted pursuant to paragraph (1).

Defence on contravention

(1)   In any proceedings against an employer for an offence under regulation 6(2), it shall be a defence for that employer to prove that -

(a)   he neither knew nor had reasonable cause to believe that he had carried out or might be required to carry out work subject to notification under that paragraph; and

(b)   in a case where he discovered that he had carried out or was carrying out work subject to notification under that paragraph, he had forthwith notified the Executive of the information required by that paragraph.

(2)   In any proceedings against an employer for an offence under regulation 7, it shall be a defence for that employer to prove that -

(a)   he neither knew nor had reasonable cause to believe that he had commenced a new activity involving work with ionising radiation; and

(b)   in a case where he had discovered that he had commenced a new activity involving work with ionising radiation, he had as soon as practicable made an assessment as required by the said regulation 7.

(3)   In any proceedings against an employer for an offence under regulation 27(2) it shall be a defence for that employer to prove that -

(a)   he had received and reasonably relied on a written undertaking from the supplier of the article concerned that it complied with the requirements of that paragraph; and

(b)   he had complied with the requirements of paragraph (3) of that regulation.

(4)   In any proceedings against an employer of an outside worker for a breach of a duty under these Regulations it shall be a defence for that employer to show that -

(a)   he had entered into a contract in writing with the employer who had designated an area as a controlled area and in which the outside worker was working or was to work for that employer to perform that duty on his behalf; and

(b)   the breach of duty was a result of the failure of the employer referred to in sub-paragraph (a) above to fulfil that contract.

(5)   In any proceedings against any employer who has designated a controlled area in which any outside worker is working or is to work for a breach of a duty under these Regulations it shall be a defence for that employer to show that -

(a)   he had entered into a contract in writing with the employer of an outside worker for that employer to perform that duty on his behalf; and

(b) the breach of duty was a result of the failure of the employer referred to in sub-paragraph (a) above to fulfil that contract.

(6) The person charged shall not, without leave of the court, be entitled to rely on the defence referred to in paragraph (4) or (5) unless, within a period ending seven clear days before the hearing, he has served on the prosecutor a notice in writing that he intends to rely on the defence and this notice shall be accompanied by a copy of the contract on which he intends to rely and, if that contract is not in English, an accurate translation of that contract into English.

(7) For the purposes of enabling the other party to be charged with and convicted of an offence by virtue of section 36 of the Health and Safety at Work etc. Act 1974, a person who establishes a defence under this regulation shall nevertheless be treated for the purposes of that section as having committed the offence.

Exemption certificates

(1) Subject to paragraph (2), the Executive may, by a certificate in writing, exempt -

(a) any person or class of persons;

(b) any premises or class of premises; or

(c) any equipment, apparatus or substance or class of equipment, apparatus or substance,

from any requirement or prohibition imposed by these Regulations and any such exemption may be granted subject to conditions and to a limit of time and may be revoked by a certificate in writing at any time.

(2) The Executive shall not grant an exemption unless, having regard to the circumstances of the case and in particular to -

(a) the conditions, if any, which it proposes to attach to the exemption; and

(b) any other requirements imposed by or under any enactments which apply to the case,

it is satisfied that -

(c) the health and safety of persons who are likely to be affected by the exemption will not be prejudiced in consequence of it; and

(d) compliance with the fundamental radiation protection provisions underlying regulations 8(1) and (2)(a), 11, 12(1), 16(1) and (3), 19(1), 20(1), 21(1), 24(2) and 32(1) will be achieved.

Extension outside Great Britain

(1) Subject to paragraph (2), these Regulations shall apply to any work outside Great Britain to which sections 1 to 59 and 80 to 82 of the Health and Safety at Work etc. Act 1974 apply by virtue of the Health and Safety at Work etc. Act 1974 (Application outside Great Britain) Order 1995[a] as they apply to work within Great Britain.

(a) SI 1995/263.

(2)   For the purposes of paragraph (1), in any case where it is not reasonably practicable for an employer to comply with the requirements of these Regulations in so far as they relate to functions being performed by an appointed doctor or employment medical adviser or by an approved dosimetry service, it shall be sufficient compliance with any such requirements if the employer makes arrangements affording an equivalent standard of protection for his employees and those arrangements are set out in local rules.

Transitional provisions

(1)   Where on or before 26th February 2000 an employer commences for the first time work which is required to be notified under regulation 6(2), it shall be sufficient compliance with that regulation if the employer notifies the Executive and notifies the required particulars before 29th January 2000.

(2)   A contingency plan made pursuant to the requirements of regulation 27 of the Ionising Radiations Regulations 1985[a] and which complied with that regulation immediately before the coming into force of these Regulations shall, for the purposes of regulation 12, be treated as if made pursuant to paragraph (1) of that regulation.

(3)   A certificate of approval granted by the Executive in respect of an approved dosimetry service under regulation 15 of the Ionising Radiations Regulations 1985 and which is valid immediately before the date of the coming into force of these Regulations, shall continue in force and shall be treated as if it had been granted under regulation 35 of these Regulations.

(4)   A radiation passbook approved for the purposes of the Ionising Radiations (Outside Workers) Regulations 1993[b] and issued prior to 30th April 2000 in respect of an outside worker employed by an employer in Great Britain and which was at that date valid shall remain valid for such time as the worker to whom the passbook relates continues to be employed by the same employer.

(5)   A doctor appointed in writing by the Executive prior to the coming into force of these Regulations for the purposes of the Ionising Radiations Regulations 1985 shall, until such time as the period specified in the appointment expires or the appointment is revoked, be deemed to have been appointed for the purposes of these Regulations.

(6)   Until 31st December 2004, an individual who or body which had before the coming into force of these Regulations been appointed by an employer as a radiation protection adviser for the purposes of the Ionising Radiations Regulations 1985 shall be deemed to meet the criteria of competence specified by the Executive for such advisers under these Regulations.

(7)   A health record which was created prior to the coming into force of these Regulations pursuant to a requirement of the Ionising Radiations Regulations 1985 shall remain valid for a period of 12 months from the date of the last entry made in it or for such shorter period as may have been specified in that record for the validity of the last entry by an appointed doctor or employment medical adviser under those Regulations, and such record shall for that period be deemed to have been kept for the purposes of regulation 24(3).

(8)   A certificate of exemption issued by the Executive pursuant to paragraph (6) of regulation 27 of the Ionising Radiations Regulations 1985 and which is valid at the coming into force of these Regulations shall continue in force until such time as it is revoked by the Executive, save that the exemption from the requirements of regulation 7 of the said 1985 Regulations shall be deemed to be an exemption from the requirements of regulation 11 of these Regulations.

(9)   Where the Executive has reasonable cause to believe that the dose received by an employee was much greater or much less than that shown in his dose record (such record having been made and maintained in accordance with regulation 13 of the Ionising Radiations Regulations 1985) the Executive may, until 30th April 2000, approve a special entry into the dose record and in such a case the employer shall arrange for the appropriate approved dosimetry service to enter the special entry in that dose record and shall give a copy of the amended dose record to the employee to whom it relates.

(a) SI 1985/1333.
(b) SI 1993/2379.

Regulation 40

Modifications relating to the Ministry of Defence etc

(1)   In this regulation, any reference to -

(a)   "visiting forces" is a reference to visiting forces within the meaning of any provision of Part 1 of the Visiting Forces Act 1952[a]; and

(b)   "headquarters or organisation" is a reference to a headquarters or organisation designated for the purposes of the International Headquarters and Defence Organisations Act 1964[b].

(2)   The Secretary of State for Defence may, in the interests of national security, by a certificate in writing exempt -

(a)   Her Majesty's Forces;

(b)   visiting forces;

(c)   any member of a visiting force working in or attached to any headquarters or organisation; or

(d)   any person engaged in work with ionising radiation for, or on behalf of, the Secretary of State for Defence,

from all or any of the requirements or prohibitions imposed by these Regulations and any such exemption may be granted subject to conditions and to a limit of time and may be revoked at any time by a certificate in writing, except that, where any such exemption is granted, suitable arrangements shall be made for the assessment and recording of doses of ionising radiation received by persons to whom the exemption relates.

(a) 1952 c.67.
(b) 1964 c.5.

(3) Sub-paragraph (i) of regulation 21(3) shall not apply in relation to a practice carried out -

(a) by or on behalf of the Secretary of State for Defence;

(b) by a visiting force; or

(c) by any member of a visiting force in or attached to any headquarters or organisation.

(4) Regulations 5 and 6 shall not apply in relation to work carried out by visiting forces or any headquarters or organisation on premises under the control of such visiting force, headquarters or organisation, as the case may be, or on premises under the control of the Secretary of State for Defence.

(5) The requirements of regulation 6 to notify the particulars specified in sub-paragraphs (d) and (e) of Schedule 2 or any of the particulars specified in Schedule 3 shall not have effect in any case where the Secretary of State for Defence decides that to do so would be against the interests of national security or where suitable alternative arrangements have been agreed with the Executive.

(6) Regulation 6(4) shall not apply to an employer in relation to work with ionising radiation undertaken for or on behalf of the Secretary of State for Defence, visiting forces or any headquarters or organisation.

(7) Regulations 22(6), (7) and (8) and regulation 24(9) shall not apply in relation to visiting forces or any member of a visiting force working in or attached to any headquarters or organisation.

(8) In regulation 25(1) the requirement to notify the Executive of a suspected overexposure and the results of the consequent investigation and assessment shall not apply in relation to the exposure of -

(a) a member of a visiting force; or

(b) a member of a visiting force working in or attached to a headquarters or organisation.

Modification, revocation and saving

(1) The enactments referred to in Schedule 9 shall be modified in accordance with the provisions of that Schedule.

(2) Subject to paragraph (3), the following enactments are hereby revoked -

(a) the Ionising Radiations Regulations 1985[a];

(b) the Ionising Radiations (Outside Workers) Regulations 1993[b];

(c) Part VI of Schedule 2 to the Personal Protective Equipment at Work Regulations 1992[c].

(a) SI 1985/1333.
(b) SI 1993/2379.
(c) SI 1992/2966.

135

(3)   Regulation 26 (Special hazard assessment) of the Ionising Radiations Regulations 1985 (in this paragraph referred to as "the 1985 Regulations") shall continue in force and, in respect of any employer subject to the said regulation 26, the following provisions shall also continue in force -

(a)   paragraphs (1) to (3), (4)(b) and (c) and (5) of regulation 27 (Contingency plans) with the modification that -

    (i)   in paragraph (1), the reference to regulation 25(1) of the 1985 Regulations shall be treated as a reference to regulation 7(1) or (2) of these Regulations;

    (ii)   in paragraph (1)(b), the reference to regulation 8(1) of Schedule 6 to the 1985 Regulations shall be treated as a reference to regulation 16 of these Regulations;

    (iii)   in paragraph (4)(b), the reference to regulation 13(2) of the 1985 Regulations shall be treated as a reference to regulation 21(2) of these Regulations;

(b)   any other provisions of the 1985 Regulations in so far as is necessary to give effect to the provisions specified in this paragraph.

(4)   Every register, certificate or record which was required to be kept in pursuance of any regulation revoked by paragraph (2) shall, notwithstanding that paragraph, be kept in the same manner and for the same period as if these Regulations had not been made, except that the Executive may approve the keeping of records at a place or in a form other than the place where, or the form in which, records were required to be kept under the regulation so revoked.

Work not required to be notified under regulation 6

Regulations 6(1) and 13(3)

1.   Work with ionising radiation shall not be required to be notified in accordance with regulation 6 when the only such work being carried out is in one or more of the following categories -

(a)   where the concentration of activity per unit mass of a radioactive substance does not exceed the concentration specified in column 2 of Part I of Schedule 8;

(b)   where the quantity of radioactive substance involved does not exceed the quantity specified in column 3 of Part I of Schedule 8;

(c)   where apparatus contains radioactive substances in a quantity exceeding the values specified in sub-paragraphs (a) and (b) above provided that -

(i)   the apparatus is of a type approved by the Executive;

(ii)   the apparatus is constructed in the form of a sealed source;

(iii)   the apparatus does not under normal operating conditions cause a dose rate of more than $1\mu Svh^{-1}$ at a distance of 0.1m from any accessible surface; and

(iv)   conditions for the disposal of the apparatus have been specified by the appropriate Agency;

(d)   the operation of any electrical apparatus to which these Regulations apply other than apparatus referred to in sub-paragraph (e) below provided that -

(i)   the apparatus is of a type approved by the Executive; and

(ii)   the apparatus does not under normal operating conditions cause a dose rate of more than $1\mu Svh^{-1}$ at a distance of 0.1m from any accessible surface;

(e)   the operation of -

(i)   any cathode ray tube intended for the display of visual images; or

(ii)   any other electrical apparatus operating at a potential difference not exceeding 30kV,

provided that the operation of the tube or apparatus does not under normal operating conditions cause a dose rate of more than $1\mu Svh^{-1}$ at a distance of 0.1m from any accessible surface;

(f)   where the work involves material contaminated with radioactive substances resulting from authorised releases which the appropriate Agency has declared not to be subject to further control.

2.   In this Schedule, "the appropriate Agency" has the meaning assigned to it by section 47(1) of the Radioactive Substances Act 1993.[a]

(a) 1993 c. 12; section 47 was amended by the Environment Act 1995 (c.25), Schedule 22, paragraph 227.

Particulars to be provided in a notification under regulation 6(2)

Regulation 6(2)

The following particulars shall be given in a notification under regulation 6(2) -

(a)   the name and address of the employer and a contact telephone or fax number or electronic mail address;

(b)   the address of the premises where or from where the work activity is to be carried out and a telephone or fax number or electronic mail address at such premises;

(c)   the nature of the business of the employer;

(d)   into which of the following categories the source or sources of ionising radiation fall -

(i)   sealed source;

(ii)   unsealed radioactive substance;

(iii) electrical equipment;

(iv)   an atmosphere containing the short-lived daughters of radon 222;

(e)   whether or not any source is to be used at premises other than the address given at sub-paragraph (b) above; and

(f)   dates of notification and commencement of the work activity.

Additional particulars that the Executive may require

Regulation 6(3)

The following additional particulars may be required under regulation 6(3) -

(a)   a description of the work with ionising radiation;

(b)   particulars of the source or sources of ionising radiation including the type of electrical equipment used or operated and the nature of any radioactive substance;

(c)   the quantities of any radioactive substance involved in the work;

(d)   the identity of any person engaged in the work;

(e)   the date of commencement and the duration of any period over which the work is carried on;

(f)   the location and description of any premises at which the work is carried out on each occasion that it is so carried out;

(g)   the date of termination of the work;

(h)   further information on any of the particulars listed in Schedule 2.

Schedule 4

Dose limits

Part I Classes of persons to whom dose limits apply

Regulation 11

Employees of 18 years of age or above

1   For the purposes of regulation 11(1), the limit on effective dose for any employee of 18 years or above shall be 20 mSv in any calendar year.

2   Without prejudice to paragraph 1 -

(a)   the limit on equivalent dose for the lens of the eye shall be 150 mSv in a calendar year;

(b)   the limit on equivalent dose for the skin shall be 500 mSv in a calendar year as applied to the dose averaged over any area of $1cm^2$ regardless of the area exposed;

(c)   the limit on equivalent dose for the hands, forearms, feet and ankles shall be 500 mSv in a calendar year.

Trainees aged under 18 years

3   For the purposes of regulation 11(1), the limit on effective dose for any trainee under 18 years of age shall be 6 mSv in any calendar year.

4   Without prejudice to paragraph 3 -

(a)   the limit on equivalent dose for the lens of the eye shall be 50 mSv in a calendar year;

(b)   the limit on equivalent dose for the skin shall be 150 mSv in a calendar year as applied to the dose averaged over any area of $1cm^2$ regardless of the area exposed;

(c)   the limit on equivalent dose for the hands, forearms, feet and ankles shall be 150 mSv in a calendar year.

Women of reproductive capacity

5   Without prejudice to paragraphs 1 and 3, the limit on equivalent dose for the abdomen of a woman of reproductive capacity who is at work, being the equivalent dose from external radiation resulting from exposure to ionising radiation averaged throughout the abdomen, shall be 13 mSv in any consecutive period of three months.

Other persons

6   Subject to paragraph 7, for the purposes of regulation 11(1) the limit on effective dose for any person other than an employee or trainee, including any person below the age of 16, shall be 1 mSv in any calendar year.

139

7   Paragraph 6 shall not apply in relation to any person (not being a comforter or carer) who may be exposed to ionising radiation resulting from the medical exposure of another and in such a case the limit on effective dose for any such person shall be 5 mSv in any period of 5 consecutive calendar years.

8   Without prejudice to paragraphs 6 and 7 -

(a)   the limit on equivalent dose for the lens of the eye shall be 15 mSv in any calendar year;

(b)   the limit on equivalent dose for the skin shall be 50 mSv in any calendar year averaged over any 1 cm^2 area regardless of the area exposed;

(c)   the limit on equivalent dose for the hands, forearms, feet and ankles shall be 50 mSv in a calendar year.

Part II

9   For the purposes of regulation 11(2), the limit on effective dose for employees of 18 years or above shall be 100 mSv in any period of five consecutive calendar years subject to a maximum effective dose of 50 mSv in any single calendar year.

10   Without prejudice to paragraph 9 -

(a)   the limit on equivalent dose for the lens of the eye shall be 150 mSv in a calendar year;

(b)   the limit on equivalent dose for the skin shall be 500 mSv in a calendar year as applied to the dose averaged over any area of 1cm^2 regardless of the area exposed;

(c)   the limit on equivalent dose for the hands, forearms, feet and ankles shall be 500 mSv in a calendar year.

11   Without prejudice to paragraph 9, the limit on equivalent dose for the abdomen of a woman of reproductive capacity who is at work, being the equivalent dose from external radiation resulting from exposure to ionising radiation averaged throughout the abdomen, shall be 13 mSv in any consecutive period of three months.

12   The employer shall ensure that any employee in respect of whom regulation 11(2) applies is not exposed to ionising radiation to an extent that any dose limit specified in paragraphs 9 to 11 is exceeded.

13   An employer shall not put into effect a system of dose limitation in pursuance of regulation 11(2) unless -

(a)   the radiation protection adviser and any employees who are affected have been consulted;

(b)   any employees affected and the approved dosimetry service have been informed in writing of the decision and of the reasons for that decision; and

(c) notice has been given to the Executive at least 28 days (or such shorter period as the Executive may allow) before the decision is put into effect giving the reasons for the decision.

14 Where there is reasonable cause to believe that any employee has been exposed to an effective dose greater than 20 mSv in any calendar year, the employer shall, as soon as is practicable -

(a) undertake an investigation into the circumstances of the exposure for the purpose of determining whether the dose limit referred to in paragraph 9 is likely to be complied with; and

(b) notify the Executive of that suspected exposure.

15 An employer shall review the decision to put into effect a system of dose limitation pursuant to regulation 11(2) at appropriate intervals and in any event not less than once every five years.

16 Where as a result of a review undertaken pursuant to paragraph 15 an employer proposes to revert to a system of annual dose limitation pursuant to regulation 11(1), the provisions of paragraph 13 shall apply as if the reference in that paragraph to regulation 11(2) was a reference to regulation 11(1).

17 Where an employer puts into effect a system of dose limitation in pursuance of regulation 11(2), he shall record the reasons for that decision and shall ensure that the record is preserved for a period of 50 years from the date of its making.

18 In any case where -

(a) the dose limits specified in paragraph 9 are being applied by a radiation employer in respect of an employee; and

(b) the Executive is not satisfied that it is impracticable for that employee to be subject to the dose limit specified in paragraph 1 of Part I of this Schedule,

the Executive may require the employer to apply the dose limit specified in paragraph 1 of Part I with effect from such time as the Executive may consider appropriate having regard to the interests of the employee concerned.

19 In any case where, as a result of a review undertaken pursuant to paragraph 15, an employer proposes to revert to an annual dose limitation pursuant to regulation 11(2), the Executive may require the employer to defer the implementation of that decision to such time as the Executive may consider appropriate having regard to the interests of the employee concerned.

20 Any person who is aggrieved by the decision of the Executive taken pursuant to paragraphs 18 or 19 may appeal to the Secretary of State.

21 Sub-sections (2) to (6) of section 44 of the 1974 Act shall apply for the purposes of paragraph 20 as they apply to an appeal under section 44(1) of that Act.

22 The Health and Safety Licensing Appeals (Hearings Procedure) Rules 1974,[a] as respects England and Wales, and the Health and Safety Licensing Appeals (Hearing Procedure) (Scotland) Rules 1974[b], as respects Scotland, shall apply to an appeal under paragraph 20 as they apply to an appeal under sub-section (1) of the said section 44, but with the modification that references to a licensing authority are to be read as references to the Executive.

(a) SI 1974/2040.
(b) SI 1974/2068.

Matters in respect of which a radiation protection adviser must be consulted by a radiation employer

Regulation 13(1)

1   The implementation of requirements as to controlled and supervised areas.

2   The prior examination of plans for installations and the acceptance into service of new or modified sources of ionising radiation in relation to any engineering controls, design features, safety features and warning devices provided to restrict exposure to ionising radiation.

3   The regular calibration of equipment provided for monitoring levels of ionising radiation and the regular checking that such equipment is serviceable and correctly used.

4   The periodic examination and testing of engineering controls, design features, safety features and warning devices and regular checking of systems of work provided to restrict exposure to ionising radiation.

Particulars to be entered in the radiation passbook

Regulation 21(5)

1   Individual serial number of the passbook.

2   A statement that the passbook has been approved by the Executive for the purpose of these Regulations.

3   Date of issue of the passbook by the approved dosimetry service.

4   The name, telephone number and mark of endorsement of the issuing approved dosimetry service.

5   The name, address, telephone and telex/fax number of the employer.

6   Full name (surname, forenames), date of birth, gender and national insurance number of the outside worker to whom the passbook has been issued.

7   Date of the last medical review of the outside worker and the relevant classification in the health record maintained under regulation 24 as fit, fit subject to conditions (which shall be specified) or unfit.

8   The relevant dose limits applicable to the outside worker to whom the passbook has been issued.

9   The cumulative dose assessment in mSv for the year to date for the outside worker, external (whole body, organ or tissue) and/or internal as appropriate and the date of the end of the last assessment period.

10   In respect of services performed by the outside worker -

(a)   the name and address of the employer responsible for the controlled area;

(b) the period covered by the performance of the services;

(c) estimated dose information, which shall be, as appropriate -

    (i) an estimate of any whole body effective dose in mSv received by the outside worker;

    (ii) in the event of non-uniform exposure, an estimate of the equivalent dose in mSv to organs and tissues as appropriate; and

    (iii) in the event of internal contamination, an estimate of the activity taken in or the committed dose.

Schedule 7

Particulars to be contained in a health record

Regulation 24(3)

The following particulars shall be contained in a health record made for the purposes of regulation 24(3) -

(a) the employee's -

    (i) full name;

    (ii) sex;

    (iii) date of birth;

    (iv) permanent address; and

    (v) National Insurance number;

(b) the date of the employee's commencement as a classified person in present employment;

(c) the nature of the employee's employment;

(d) in the case of a female employee, a statement as to whether she is likely to receive in any consecutive period of three months an equivalent dose of ionising radiation for the abdomen exceeding 13 mSv;

(e) the date of last medical examination or health review carried out in respect of the employee;

(f) the type of the last medical examination or health review carried out in respect of the employee;

(g) a statement by the appointed doctor or employment medical adviser made as a result of the last medical examination or health review carried out in respect of the employee classifying the employee as fit, fit subject to conditions (which should be specified) or unfit;

(h) in the case of a female employee in respect of whom a statement has been made under paragraph (d) to the effect that she is likely to receive in any consecutive period of three months an equivalent dose of ionising radiation for the abdomen exceeding 13 mSv, a statement by the appointed doctor or employment medical adviser certifying whether in his professional opinion the employee should be subject to the additional dose limit specified in paragraphs 5 and 11 of Schedule 4;

(i)   in relation to each medical examination and health review, the name and signature of the appointed doctor or employment medical adviser;

(j)   the name and address of the approved dosimetry service with whom arrangements have been made for maintaining the dose record in accordance with regulation 21.

Schedule 8 — Quantities and concentrations of radionuclides

Regulation 2(4) and 30(1) and (2) and Schedule 1

Part I Table of radionuclides

1 Radionuclide name, symbol, isotope	2 Concentration for notification. Regulation 6 and Schedule 1 (Bq/g)	3 Quantity for notification. Regulation 6 and Schedule 1 (Bq)	4 Quantity for notification of occurrences. Regulation 30(1) (Bq)	5 Quantity for notification of occurrences. Regulation 30(3) (Bq)
Hydrogen Tritiated Compounds Elemental	$1\ 10^6$ $1\ 10^6$	$1\ 10^9$ $1\ 10^9$	$1\ 10^{12}$ $1\ 10^{13}$	$1\ 10^{10}$ $1\ 10^{10}$
Beryllium Be-7 Be-10	$1\ 10^3$ $1\ 10^4$	$1\ 10^7$ $1\ 10^6$	$1\ 10^{12}$ $1\ 10^{10}$	$1\ 10^8$ $1\ 10^7$
Carbon C-11 C-11 monoxide C-11 dioxide C-14 C-14 monoxide C-14 dioxide	$1\ 10^1$ $1\ 10^1$ $1\ 10^1$ $1\ 10^4$ $1\ 10^8$ $1\ 10^7$	$1\ 10^6$ $1\ 10^9$ $1\ 10^9$ $1\ 10^7$ $1\ 10^{11}$ $1\ 10^{11}$	$1\ 10^{13}$ $1\ 10^{12}$ $1\ 10^{12}$ $1\ 10^{11}$ $1\ 10^{14}$ $1\ 10^{13}$	$1\ 10^7$ $1\ 10^{10}$ $1\ 10^{10}$ $1\ 10^8$ $1\ 10^{12}$ $1\ 10^{12}$
Nitrogen N-13	$1\ 10^2$	$1\ 10^9$	$1\ 10^9$	
Oxygen O-15	$1\ 10^2$	$1\ 10^9$	$1\ 10^{10}$	

1 Radionuclide name, symbol, isotope	2 Concentration for notification. Regulation 6 and Schedule 1 (Bq/g)	3 Quantity for notification. Regulation 6 and Schedule 1 (Bq)	4 Quantity for notification of occurrences. Regulation 30(1) (Bq)	5 Quantity for notification of occurrences. Regulation 30(3) (Bq)
Fluorine F-18	$1\ 10^1$	$1\ 10^6$	$1\ 10^{13}$	$1\ 10^7$
Neon Ne-19	$1\ 10^2$	$1\ 10^9$	$1\ 10^9$	
Sodium Na-22 Na-24	$1\ 10^1$ $1\ 10^1$	$1\ 10^6$ $1\ 10^5$	$1\ 10^{10}$ $1\ 10^{11}$	$1\ 10^7$ $1\ 10^6$
Magnesium Mg-28+	$1\ 10^1$	$1\ 10^5$	$1\ 10^{11}$	$1\ 10^6$
Aluminium Al-26	$1\ 10^1$	$1\ 10^5$	$1\ 10^{10}$	$1\ 10^6$
Silicon Si-31 Si-32	$1\ 10^3$ $1\ 10^3$	$1\ 10^6$ $1\ 10^6$	$1\ 10^{13}$ $1\ 10^9$	$1\ 10^7$ $1\ 10^7$
Phosphorus P-32 P-33	$1\ 10^3$ $1\ 10^5$	$1\ 10^5$ $1\ 10^8$	$1\ 10^{10}$ $1\ 10^{11}$	$1\ 10^6$ $1\ 10^9$
Sulphur S-35 S-35 (organic) S-35 Vapour	$1\ 10^5$ $1\ 10^5$ $1\ 10^6$	$1\ 10^8$ $1\ 10^8$ $1\ 10^9$	$1\ 10^{11}$ $1\ 10^{12}$ $1\ 10^{12}$	$1\ 10^9$ $1\ 10^9$
Chlorine Cl-36 Cl-38 Cl-39	$1\ 10^4$ $1\ 10^1$ $1\ 10^1$	$1\ 10^6$ $1\ 10^5$ $1\ 10^5$	$1\ 10^{10}$ $1\ 10^{13}$ $1\ 10^{13}$	$1\ 10^7$ $1\ 10^6$ $1\ 10^6$
Argon Ar-37 Ar-39 Ar-41	$1\ 10^6$ $1\ 10^7$ $1\ 10^2$	$1\ 10^8$ $1\ 10^4$ $1\ 10^9$	$1\ 10^{13}$ $1\ 10^{12}$ $1\ 10^9$	
Potassium K-40 K-42 K-43 K-44 K-45	$1\ 10^2$ $1\ 10^2$ $1\ 10^1$ $1\ 10^1$ $1\ 10^1$	$1\ 10^6$ $1\ 10^6$ $1\ 10^6$ $1\ 10^5$ $1\ 10^5$	$1\ 10^{10}$ $1\ 10^{12}$ $1\ 10^{11}$ $1\ 10^{13}$ $1\ 10^{13}$	$1\ 10^7$ $1\ 10^7$ $1\ 10^7$ $1\ 10^6$ $1\ 10^6$
Calcium Ca-41 Ca-45 Ca-47	$1\ 10^5$ $1\ 10^4$ $1\ 10^1$	$1\ 10^7$ $1\ 10^7$ $1\ 10^6$	$1\ 10^{12}$ $1\ 10^{10}$ $1\ 10^{11}$	$1\ 10^8$ $1\ 10^8$ $1\ 10^7$

1 Radionuclide name, symbol, isotope	2 Concentration for notification. Regulation 6 and Schedule 1 (Bq/g)	3 Quantity for notification. Regulation 6 and Schedule 1 (Bq)	4 Quantity for notification of occurrences. Regulation 30(1) (Bq)	5 Quantity for notification of occurrences. Regulation 30(3) (Bq)
Scandium				
Sc-43	$1\ 10^1$	$1\ 10^6$	$1\ 10^{12}$	$1\ 10^7$
Sc-44	$1\ 10^1$	$1\ 10^5$	$1\ 10^{12}$	$1\ 10^6$
Sc-44m	$1\ 10^2$	$1\ 10^7$	$1\ 10^{11}$	$1\ 10^8$
Sc-46	$1\ 10^1$	$1\ 10^6$	$1\ 10^{10}$	$1\ 10^7$
Sc-47	$1\ 10^2$	$1\ 10^6$	$1\ 10^{11}$	$1\ 10^7$
Sc-48	$1\ 10^1$	$1\ 10^5$	$1\ 10^{11}$	$1\ 10^6$
Sc-49	$1\ 10^3$	$1\ 10^5$	$1\ 10^{14}$	$1\ 10^6$
Titanium				
Ti-44 +	$1\ 10^1$	$1\ 10^5$	$1\ 10^9$	$1\ 10^6$
Ti-45	$1\ 10^1$	$1\ 10^6$	$1\ 10^{12}$	$1\ 10^7$
Vanadium				
V-47	$1\ 10^1$	$1\ 10^5$	$1\ 10^{13}$	$1\ 10^6$
V-48	$1\ 10^1$	$1\ 10^5$	$1\ 10^{10}$	$1\ 10^6$
V-49	$1\ 10^4$	$1\ 10^7$	$1\ 10^{12}$	$1\ 10^8$
Chromium				
Cr-48	$1\ 10^2$	$1\ 10^6$	$1\ 10^{12}$	$1\ 10^7$
Cr-49	$1\ 10^1$	$1\ 10^6$	$1\ 10^{13}$	$1\ 10^7$
Cr-51	$1\ 10^3$	$1\ 10^7$	$1\ 10^{12}$	$1\ 10^8$
Manganese				
Mn-51	$1\ 10^1$	$1\ 10^5$	$1\ 10^{13}$	$1\ 10^6$
Mn-52	$1\ 10^1$	$1\ 10^5$	$1\ 10^{10}$	$1\ 10^6$
Mn-52m	$1\ 10^1$	$1\ 10^5$	$1\ 10^{13}$	$1\ 10^6$
Mn-53	$1\ 10^4$	$1\ 10^9$	$1\ 10^{12}$	$1\ 10^{10}$
Mn-54	$1\ 10^1$	$1\ 10^6$	$1\ 10^{11}$	$1\ 10^7$
Mn-56	$1\ 10^1$	$1\ 10^5$	$1\ 10^{12}$	$1\ 10^6$
Iron				
Fe-52	$1\ 10^1$	$1\ 10^6$	$1\ 10^{12}$	$1\ 10^7$
Fe-55	$1\ 10^4$	$1\ 10^6$	$1\ 10^{11}$	$1\ 10^7$
Fe-59	$1\ 10^1$	$1\ 10^6$	$1\ 10^{10}$	$1\ 10^7$
Fe-60+	$1\ 10^2$	$1\ 10^5$	$1\ 10^8$	$1\ 10^6$
Cobalt				
Co-55	$1\ 10^1$	$1\ 10^6$	$1\ 10^{11}$	$1\ 10^7$
Co-56	$1\ 10^1$	$1\ 10^5$	$1\ 10^{10}$	$1\ 10^6$
Co-57	$1\ 10^2$	$1\ 10^6$	$1\ 10^{11}$	$1\ 10^7$
Co-58	$1\ 10^1$	$1\ 10^6$	$1\ 10^{10}$	$1\ 10^7$
Co-58m	$1\ 10^4$	$1\ 10^7$	$1\ 10^{13}$	$1\ 10^8$
Co-60	$1\ 10^1$	$1\ 10^5$	$1\ 10^{10}$	$1\ 10^6$
Co-60m	$1\ 10^3$	$1\ 10^6$	$1\ 10^{16}$	$1\ 10^7$
Co-61	$1\ 10^2$	$1\ 10^6$	$1\ 10^{13}$	$1\ 10^7$
Co-62m	$1\ 10^1$	$1\ 10^5$	$1\ 10^{13}$	$1\ 10^6$

1 Radionuclide name, symbol, isotope	2 Concentration for notification. Regulation 6 and Schedule 1 (Bq/g)	3 Quantity for notification. Regulation 6 and Schedule 1 (Bq)	4 Quantity for notification of occurrences. Regulation 30(1) (Bq)	5 Quantity for notification of occurrences. Regulation 30(3) (Bq)
Nickel				
Ni-56	$1\ 10^1$	$1\ 10^6$	$1\ 10^{11}$	$1\ 10^7$
Ni-57	$1\ 10^1$	$1\ 10^6$	$1\ 10^{11}$	$1\ 10^7$
Ni-59	$1\ 10^4$	$1\ 10^8$	$1\ 10^{11}$	$1\ 10^9$
Ni-63	$1\ 10^5$	$1\ 10^8$	$1\ 10^{11}$	$1\ 10^9$
Ni-65	$1\ 10^1$	$1\ 10^6$	$1\ 10^{13}$	$1\ 10^7$
Ni-66	$1\ 10^4$	$1\ 10^7$	$1\ 10^{11}$	$1\ 10^8$
Copper				
Cu-60	$1\ 10^1$	$1\ 10^5$	$1\ 10^{13}$	$1\ 10^6$
Cu-61	$1\ 10^1$	$1\ 10^6$	$1\ 10^{12}$	$1\ 10^7$
Cu-64	$1\ 10^2$	$1\ 10^6$	$1\ 10^{12}$	$1\ 10^7$
Cu-67	$1\ 10^2$	$1\ 10^6$	$1\ 10^{11}$	$1\ 10^7$
Zinc				
Zn-62	$1\ 10^2$	$1\ 10^6$	$1\ 10^{12}$	$1\ 10^7$
Zn-63	$1\ 10^1$	$1\ 10^5$	$1\ 10^{13}$	$1\ 10^6$
Zn-65	$1\ 10^1$	$1\ 10^6$	$1\ 10^{10}$	$1\ 10^7$
Zn-69	$1\ 10^4$	$1\ 10^6$	$1\ 10^{14}$	$1\ 10^7$
Zn-69m	$1\ 10^2$	$1\ 10^6$	$1\ 10^{12}$	$1\ 10^7$
Zn-71m	$1\ 10^1$	$1\ 10^6$	$1\ 10^{12}$	$1\ 10^7$
Zn-72	$1\ 10^2$	$1\ 10^6$	$1\ 10^{11}$	$1\ 10^7$
Gallium				
Ga-65	$1\ 10^1$	$1\ 10^5$	$1\ 10^{13}$	$1\ 10^6$
Ga-66	$1\ 10^1$	$1\ 10^5$	$1\ 10^{11}$	$1\ 10^6$
Ga-67	$1\ 10^2$	$1\ 10^6$	$1\ 10^{11}$	$1\ 10^7$
Ga-68	$1\ 10^1$	$1\ 10^5$	$1\ 10^{13}$	$1\ 10^6$
Ga-70	$1\ 10^3$	$1\ 10^6$	$1\ 10^{14}$	$1\ 10^7$
Ga-72	$1\ 10^1$	$1\ 10^5$	$1\ 10^{11}$	$1\ 10^6$
Ga-73	$1\ 10^2$	$1\ 10^6$	$1\ 10^{12}$	$1\ 10^7$
Germanium				
Ge-66	$1\ 10^1$	$1\ 10^6$	$1\ 10^{13}$	$1\ 10^7$
Ge-67	$1\ 10^1$	$1\ 10^5$	$1\ 10^{13}$	$1\ 10^6$
Ge-68+	$1\ 10^1$	$1\ 10^5$	$1\ 10^{10}$	$1\ 10^6$
Ge-69	$1\ 10^1$	$1\ 10^6$	$1\ 10^{11}$	$1\ 10^7$
Ge-71	$1\ 10^4$	$1\ 10^8$	$1\ 10^{13}$	$1\ 10^9$
Ge-75	$1\ 10^3$	$1\ 10^6$	$1\ 10^{14}$	$1\ 10^7$
Ge-77	$1\ 10^1$	$1\ 10^5$	$1\ 10^{12}$	$1\ 10^6$
Ge-78	$1\ 10^2$	$1\ 10^6$	$1\ 10^{13}$	$1\ 10^7$
Arsenic				
As-69	$1\ 10^1$	$1\ 10^5$	$1\ 10^{13}$	$1\ 10^6$
As-70	$1\ 10^1$	$1\ 10^5$	$1\ 10^{12}$	$1\ 10^6$
As-71	$1\ 10^1$	$1\ 10^6$	$1\ 10^{11}$	$1\ 10^7$
As-72	$1\ 10^1$	$1\ 10^5$	$1\ 10^{11}$	$1\ 10^6$
As-73	$1\ 10^3$	$1\ 10^7$	$1\ 10^{11}$	$1\ 10^8$
As-74	$1\ 10^1$	$1\ 10^6$	$1\ 10^{11}$	$1\ 10^7$
As-76	$1\ 10^2$	$1\ 10^5$	$1\ 10^{11}$	$1\ 10^6$
As-77	$1\ 10^3$	$1\ 10^6$	$1\ 10^{12}$	$1\ 10^7$
As-78	$1\ 10^1$	$1\ 10^5$	$1\ 10^{13}$	$1\ 10^6$

147

1 Radionuclide name, symbol, isotope	2 Concentration for notification. Regulation 6 and Schedule 1 (Bq/g)	3 Quantity for notification. Regulation 6 and Schedule 1 (Bq)	4 Quantity for notification of occurrences. Regulation 30(1) (Bq)	5 Quantity for notification of occurrences. Regulation 30(3) (Bq)
Selenium				
Se-70	$1 \ 10^1$	$1 \ 10^6$	$1 \ 10^{13}$	$1 \ 10^7$
Se-73	$1 \ 10^1$	$1 \ 10^6$	$1 \ 10^{12}$	$1 \ 10^7$
Se-73m	$1 \ 10^2$	$1 \ 10^6$	$1 \ 10^{14}$	$1 \ 10^7$
Se-75	$1 \ 10^2$	$1 \ 10^6$	$1 \ 10^{11}$	$1 \ 10^7$
Se-79	$1 \ 10^4$	$1 \ 10^7$	$1 \ 10^{10}$	$1 \ 10^8$
Se-81	$1 \ 10^3$	$1 \ 10^6$	$1 \ 10^{14}$	$1 \ 10^7$
Se-81m	$1 \ 10^3$	$1 \ 10^7$	$1 \ 10^{14}$	$1 \ 10^8$
Se-83	$1 \ 10^1$	$1 \ 10^5$	$1 \ 10^{13}$	$1 \ 10^6$
Bromine				
Br-74	$1 \ 10^1$	$1 \ 10^5$	$1 \ 10^{13}$	$1 \ 10^6$
Br-74m	$1 \ 10^1$	$1 \ 10^5$	$1 \ 10^{12}$	$1 \ 10^6$
Br-75	$1 \ 10^1$	$1 \ 10^6$	$1 \ 10^{13}$	$1 \ 10^7$
Br-76	$1 \ 10^1$	$1 \ 10^5$	$1 \ 10^{11}$	$1 \ 10^6$
Br-77	$1 \ 10^2$	$1 \ 10^6$	$1 \ 10^{12}$	$1 \ 10^7$
Br-80	$1 \ 10^2$	$1 \ 10^5$	$1 \ 10^{14}$	$1 \ 10^6$
Br-80m	$1 \ 10^3$	$1 \ 10^7$	$1 \ 10^{13}$	$1 \ 10^8$
Br-82	$1 \ 10^1$	$1 \ 10^6$	$1 \ 10^{11}$	$1 \ 10^7$
Br-83	$1 \ 10^3$	$1 \ 10^6$	$1 \ 10^{13}$	$1 \ 10^7$
Br-84	$1 \ 10^1$	$1 \ 10^5$	$1 \ 10^{13}$	$1 \ 10^6$
Krypton				
Kr-74	$1 \ 10^2$	$1 \ 10^9$	$1 \ 10^9$	
Kr-76	$1 \ 10^2$	$1 \ 10^9$	$1 \ 10^{10}$	
Kr-77	$1 \ 10^2$	$1 \ 10^9$	$1 \ 10^9$	
Kr-79	$1 \ 10^3$	$1 \ 10^5$	$1 \ 10^{10}$	
Kr-81	$1 \ 10^4$	$1 \ 10^7$	$1 \ 10^{11}$	
Kr-81m	$1 \ 10^3$	$1 \ 10^{10}$	$1 \ 10^{10}$	
Kr-83m	$1 \ 10^5$	$1 \ 10^{12}$	$1 \ 10^{12}$	
Kr-85	$1 \ 10^5$	$1 \ 10^4$	$1 \ 10^{12}$	
Kr-85m	$1 \ 10^3$	$1 \ 10^{10}$	$1 \ 10^{10}$	
Kr-87	$1 \ 10^2$	$1 \ 10^9$	$1 \ 10^9$	
Kr-88	$1 \ 10^2$	$1 \ 10^9$	$1 \ 10^9$	
Rubidium				
Rb-79	$1 \ 10^1$	$1 \ 10^5$	$1 \ 10^{13}$	$1 \ 10^6$
Rb-81	$1 \ 10^1$	$1 \ 10^6$	$1 \ 10^{12}$	$1 \ 10^7$
Rb-81m	$1 \ 10^3$	$1 \ 10^7$	$1 \ 10^{15}$	$1 \ 10^8$
Rb-82m	$1 \ 10^1$	$1 \ 10^6$	$1 \ 10^{12}$	$1 \ 10^7$
Rb-83+	$1 \ 10^2$	$1 \ 10^6$	$1 \ 10^{11}$	$1 \ 10^7$
Rb-84	$1 \ 10^1$	$1 \ 10^6$	$1 \ 10^{11}$	$1 \ 10^7$
Rb-86	$1 \ 10^2$	$1 \ 10^5$	$1 \ 10^{11}$	$1 \ 10^6$
Rb-87	$1 \ 10^4$	$1 \ 10^7$	$1 \ 10^{11}$	$1 \ 10^8$
Rb-88	$1 \ 10^1$	$1 \ 10^5$	$1 \ 10^{14}$	$1 \ 10^6$
Rb-89	$1 \ 10^1$	$1 \ 10^5$	$1 \ 10^{13}$	$1 \ 10^6$

1 Radionuclide name, symbol, isotope	2 Concentration for notification. Regulation 6 and Schedule 1 (Bq/g)	3 Quantity for notification. Regulation 6 and Schedule 1 (Bq)	4 Quantity for notification of occurrences. Regulation 30(1) (Bq)	5 Quantity for notification of occurrences. Regulation 30(3) (Bq)
Strontium				
Sr-80	$1\ 10^3$	$1\ 10^7$	$1\ 10^{13}$	$1\ 10^8$
Sr-81	$1\ 10^1$	$1\ 10^5$	$1\ 10^{13}$	$1\ 10^6$
Sr-82+	$1\ 10^1$	$1\ 10^5$	$1\ 10^{10}$	$1\ 10^6$
Sr-83	$1\ 10^1$	$1\ 10^6$	$1\ 10^{11}$	$1\ 10^7$
Sr-85	$1\ 10^2$	$1\ 10^6$	$1\ 10^{11}$	$1\ 10^7$
Sr-85m	$1\ 10^2$	$1\ 10^7$	$1\ 10^{13}$	$1\ 10^8$
Sr-87m	$1\ 10^2$	$1\ 10^6$	$1\ 10^{13}$	$1\ 10^7$
Sr-89	$1\ 10^3$	$1\ 10^6$	$1\ 10^{10}$	$1\ 10^7$
Sr-90+	$1\ 10^2$	$1\ 10^4$	$1\ 10^9$	$1\ 10^5$
Sr-91	$1\ 10^1$	$1\ 10^5$	$1\ 10^{12}$	$1\ 10^6$
Sr-92	$1\ 10^1$	$1\ 10^6$	$1\ 10^{12}$	$1\ 10^7$
Yttrium				
Y-86	$1\ 10^1$	$1\ 10^5$	$1\ 10^{11}$	$1\ 10^6$
Y-86m	$1\ 10^2$	$1\ 10^7$	$1\ 10^{14}$	$1\ 10^8$
Y-87+	$1\ 10^1$	$1\ 10^6$	$1\ 10^{11}$	$1\ 10^7$
Y-88	$1\ 10^1$	$1\ 10^6$	$1\ 10^{10}$	$1\ 10^7$
Y-90	$1\ 10^3$	$1\ 10^5$	$1\ 10^{11}$	$1\ 10^6$
Y-90m	$1\ 10^1$	$1\ 10^6$	$1\ 10^{12}$	$1\ 10^7$
Y-91	$1\ 10^3$	$1\ 10^6$	$1\ 10^{10}$	$1\ 10^7$
Y-91m	$1\ 10^2$	$1\ 10^6$	$1\ 10^{13}$	$1\ 10^7$
Y-92	$1\ 10^2$	$1\ 10^5$	$1\ 10^{12}$	$1\ 10^6$
Y-93	$1\ 10^2$	$1\ 10^5$	$1\ 10^{12}$	$1\ 10^6$
Y-94	$1\ 10^1$	$1\ 10^5$	$1\ 10^{13}$	$1\ 10^6$
Y-95	$1\ 10^1$	$1\ 10^5$	$1\ 10^{14}$	$1\ 10^6$
Zirconium				
Zr-86	$1\ 10^2$	$1\ 10^7$	$1\ 10^{12}$	$1\ 10^8$
Zr-88	$1\ 10^2$	$1\ 10^6$	$1\ 10^{10}$	$1\ 10^7$
Zr-89	$1\ 10^1$	$1\ 10^6$	$1\ 10^{11}$	$1\ 10^7$
Zr-93+	$1\ 10^3$	$1\ 10^7$	$1\ 10^9$	$1\ 10^8$
Zr-95	$1\ 10^1$	$1\ 10^6$	$1\ 10^{10}$	$1\ 10^7$
Zr-97+	$1\ 10^1$	$1\ 10^5$	$1\ 10^{11}$	$1\ 10^6$
Niobium				
Nb-88	$1\ 10^1$	$1\ 10^5$	$1\ 10^{13}$	$1\ 10^6$
Nb-89 (2.03 hours)	$1\ 10^1$	$1\ 10^5$	$1\ 10^{12}$	$1\ 10^6$
Nb-89 (1.01 hour)	$1\ 10^1$	$1\ 10^5$	$1\ 10^{13}$	$1\ 10^6$
Nb-90	$1\ 10^1$	$1\ 10^5$	$1\ 10^{11}$	$1\ 10^6$
Nb-93m	$1\ 10^4$	$1\ 10^7$	$1\ 10^{11}$	$1\ 10^8$
Nb-94	$1\ 10^1$	$1\ 10^6$	$1\ 10^9$	$1\ 10^7$
Nb-95	$1\ 10^1$	$1\ 10^6$	$1\ 10^{11}$	$1\ 10^7$
Nb-95m	$1\ 10^2$	$1\ 10^7$	$1\ 10^{11}$	$1\ 10^8$
Nb-96	$1\ 10^1$	$1\ 10^5$	$1\ 10^{11}$	$1\ 10^6$
Nb-97	$1\ 10^1$	$1\ 10^6$	$1\ 10^{13}$	$1\ 10^7$
Nb-98	$1\ 10^1$	$1\ 10^5$	$1\ 10^{13}$	$1\ 10^6$

1 Radionuclide name, symbol, isotope	2 Concentration for notification. Regulation 6 and Schedule 1 (Bq/g)	3 Quantity for notification. Regulation 6 and Schedule 1 (Bq)	4 Quantity for notification of occurrences. Regulation 30(1) (Bq)	5 Quantity for notification of occurrences. Regulation 30(3) (Bq)
Molybdenum				
Mo-90	$1\ 10^1$	$1\ 10^6$	$1\ 10^{12}$	$1\ 10^7$
Mo-93	$1\ 10^3$	$1\ 10^8$	$1\ 10^{11}$	$1\ 10^9$
Mo-93m	$1\ 10^1$	$1\ 10^6$	$1\ 10^{12}$	$1\ 10^7$
Mo-99	$1\ 10^2$	$1\ 10^6$	$1\ 10^{11}$	$1\ 10^7$
Mo-101	$1\ 10^1$	$1\ 10^6$	$1\ 10^{13}$	$1\ 10^7$
Technetium				
Tc-93	$1\ 10^1$	$1\ 10^6$	$1\ 10^{12}$	$1\ 10^7$
Tc-93m	$1\ 10^1$	$1\ 10^6$	$1\ 10^{13}$	$1\ 10^7$
Tc-94	$1\ 10^1$	$1\ 10^6$	$1\ 10^{12}$	$1\ 10^7$
Tc-94m	$1\ 10^1$	$1\ 10^5$	$1\ 10^{13}$	$1\ 10^6$
Tc-95	$1\ 10^1$	$1\ 10^6$	$1\ 10^{12}$	$1\ 10^7$
Tc-95m+	$1\ 10^1$	$1\ 10^6$	$1\ 10^{11}$	$1\ 10^7$
Tc-96	$1\ 10^1$	$1\ 10^6$	$1\ 10^{11}$	$1\ 10^7$
Tc-96m	$1\ 10^3$	$1\ 10^7$	$1\ 10^{14}$	$1\ 10^8$
Tc-97	$1\ 10^3$	$1\ 10^8$	$1\ 10^{12}$	$1\ 10^9$
Tc-97m	$1\ 10^3$	$1\ 10^7$	$1\ 10^{10}$	$1\ 10^8$
Tc-98	$1\ 10^1$	$1\ 10^6$	$1\ 10^{10}$	$1\ 10^7$
Tc-99	$1\ 10^4$	$1\ 10^7$	$1\ 10^{10}$	$1\ 10^8$
Tc-99m	$1\ 10^2$	$1\ 10^7$	$1\ 10^{13}$	$1\ 10^8$
Tc-101	$1\ 10^2$	$1\ 10^6$	$1\ 10^{14}$	$1\ 10^7$
Tc-104	$1\ 10^1$	$1\ 10^5$	$1\ 10^{13}$	$1\ 10^6$
Ruthenium				
Ru-94	$1\ 10^2$	$1\ 10^6$	$1\ 10^{13}$	$1\ 10^7$
Ru-97	$1\ 10^2$	$1\ 10^7$	$1\ 10^{12}$	$1\ 10^8$
Ru-103	$1\ 10^2$	$1\ 10^6$	$1\ 10^{10}$	$1\ 10^7$
Ru-105	$1\ 10^1$	$1\ 10^6$	$1\ 10^{12}$	$1\ 10^7$
Ru-106+	$1\ 10^2$	$1\ 10^5$	$1\ 10^9$	$1\ 10^6$
Rhodium				
Rh-99	$1\ 10^1$	$1\ 10^6$	$1\ 10^{11}$	$1\ 10^7$
Rh-99m	$1\ 10^1$	$1\ 10^6$	$1\ 10^{12}$	$1\ 10^7$
Rh-100	$1\ 10^1$	$1\ 10^6$	$1\ 10^{11}$	$1\ 10^7$
Rh-101	$1\ 10^2$	$1\ 10^7$	$1\ 10^{10}$	$1\ 10^8$
Rh-101m	$1\ 10^2$	$1\ 10^7$	$1\ 10^{11}$	$1\ 10^8$
Rh-102	$1\ 10^1$	$1\ 10^6$	$1\ 10^{10}$	$1\ 10^7$
Rh-102m	$1\ 10^2$	$1\ 10^6$	$1\ 10^{10}$	$1\ 10^7$
Rh-103m	$1\ 10^4$	$1\ 10^8$	$1\ 10^{15}$	$1\ 10^9$
Rh-105	$1\ 10^2$	$1\ 10^7$	$1\ 10^{12}$	$1\ 10^8$
Rh-106m	$1\ 10^1$	$1\ 10^5$	$1\ 10^{12}$	$1\ 10^6$
Rh-107	$1\ 10^2$	$1\ 10^6$	$1\ 10^{14}$	$1\ 10^7$
Palladium				
Pd-100	$1\ 10^2$	$1\ 10^7$	$1\ 10^{11}$	$1\ 10^8$
Pd-101	$1\ 10^2$	$1\ 10^6$	$1\ 10^{12}$	$1\ 10^7$
Pd-103	$1\ 10^3$	$1\ 10^8$	$1\ 10^{11}$	$1\ 10^9$
Pd-107	$1\ 10^5$	$1\ 10^8$	$1\ 10^{11}$	$1\ 10^9$
Pd-109	$1\ 10^3$	$1\ 10^6$	$1\ 10^{12}$	$1\ 10^7$

1 Radionuclide name, symbol, isotope	2 Concentration for notification. Regulation 6 and Schedule 1 (Bq/g)	3 Quantity for notification. Regulation 6 and Schedule 1 (Bq)	4 Quantity for notification of occurrences. Regulation 30(1) (Bq)	5 Quantity for notification of occurrences. Regulation 30(3) (Bq)
Silver				
Ag-102	$1\ 10^1$	$1\ 10^5$	$1\ 10^{13}$	$1\ 10^6$
Ag-103	$1\ 10^1$	$1\ 10^6$	$1\ 10^{13}$	$1\ 10^7$
Ag-104	$1\ 10^1$	$1\ 10^6$	$1\ 10^{12}$	$1\ 10^7$
Ag-104m	$1\ 10^1$	$1\ 10^6$	$1\ 10^{13}$	$1\ 10^7$
Ag-105	$1\ 10^2$	$1\ 10^6$	$1\ 10^{11}$	$1\ 10^7$
Ag-106	$1\ 10^1$	$1\ 10^6$	$1\ 10^{13}$	$1\ 10^7$
Ag-106m	$1\ 10^1$	$1\ 10^6$	$1\ 10^{10}$	$1\ 10^7$
Ag-108m+	$1\ 10^1$	$1\ 10^6$	$1\ 10^{10}$	$1\ 10^7$
Ag-110m	$1\ 10^1$	$1\ 10^6$	$1\ 10^{10}$	$1\ 10^7$
Ag-111	$1\ 10^3$	$1\ 10^6$	$1\ 10^{11}$	$1\ 10^7$
Ag-112	$1\ 10^1$	$1\ 10^5$	$1\ 10^{12}$	$1\ 10^6$
Ag-115	$1\ 10^1$	$1\ 10^5$	$1\ 10^{13}$	$1\ 10^6$
Cadmium				
Cd-104	$1\ 10^2$	$1\ 10^7$	$1\ 10^{13}$	$1\ 10^8$
Cd-107	$1\ 10^3$	$1\ 10^7$	$1\ 10^{13}$	$1\ 10^8$
Cd-109	$1\ 10^4$	$1\ 10^6$	$1\ 10^{10}$	$1\ 10^7$
Cd-113	$1\ 10^3$	$1\ 10^6$	$1\ 10^9$	$1\ 10^7$
Cd-113m	$1\ 10^3$	$1\ 10^6$	$1\ 10^9$	$1\ 10^7$
Cd-115	$1\ 10^2$	$1\ 10^6$	$1\ 10^{11}$	$1\ 10^7$
Cd-115m	$1\ 10^3$	$1\ 10^6$	$1\ 10^{10}$	$1\ 10^7$
Cd-117	$1\ 10^1$	$1\ 10^6$	$1\ 10^{12}$	$1\ 10^7$
Cd-117m	$1\ 10^1$	$1\ 10^6$	$1\ 10^{12}$	$1\ 10^7$
Indium				
In-109	$1\ 10^1$	$1\ 10^6$	$1\ 10^{12}$	$1\ 10^7$
In-110 (4.9 hours)	$1\ 10^1$	$1\ 10^6$	$1\ 10^{12}$	$1\ 10^7$
In-110 (69.1 min)	$1\ 10^1$	$1\ 10^5$	$1\ 10^{13}$	$1\ 10^6$
In-111	$1\ 10^2$	$1\ 10^6$	$1\ 10^{11}$	$1\ 10^7$
In-112	$1\ 10^2$	$1\ 10^6$	$1\ 10^{14}$	$1\ 10^7$
In-113m	$1\ 10^2$	$1\ 10^6$	$1\ 10^{13}$	$1\ 10^7$
In-114	$1\ 10^3$	$1\ 10^5$	$1\ 10^{15}$	$1\ 10^6$
In-114m	$1\ 10^2$	$1\ 10^6$	$1\ 10^{10}$	$1\ 10^7$
In-115	$1\ 10^3$	$1\ 10^5$	$1\ 10^8$	$1\ 10^6$
In-115m	$1\ 10^2$	$1\ 10^6$	$1\ 10^{13}$	$1\ 10^7$
In-116m	$1\ 10^1$	$1\ 10^5$	$1\ 10^{13}$	$1\ 10^6$
In-117	$1\ 10^1$	$1\ 10^6$	$1\ 10^{13}$	$1\ 10^7$
In-117m	$1\ 10^2$	$1\ 10^6$	$1\ 10^{13}$	$1\ 10^7$
In-119m	$1\ 10^2$	$1\ 10^5$	$1\ 10^{14}$	$1\ 10^6$
Tin				
Sn-110	$1\ 10^2$	$1\ 10^7$	$1\ 10^{12}$	$1\ 10^8$
Sn-111	$1\ 10^2$	$1\ 10^6$	$1\ 10^{13}$	$1\ 10^7$
Sn-113	$1\ 10^3$	$1\ 10^7$	$1\ 10^{11}$	$1\ 10^8$
Sn-117m	$1\ 10^2$	$1\ 10^6$	$1\ 10^{11}$	$1\ 10^7$
Sn-119m	$1\ 10^3$	$1\ 10^7$	$1\ 10^{11}$	$1\ 10^8$
Sn-121	$1\ 10^5$	$1\ 10^7$	$1\ 10^{12}$	$1\ 10^8$
Sn-121m+	$1\ 10^3$	$1\ 10^7$	$1\ 10^{10}$	$1\ 10^8$
Sn-123	$1\ 10^3$	$1\ 10^6$	$1\ 10^{10}$	$1\ 10^7$
Sn-123m	$1\ 10^2$	$1\ 10^6$	$1\ 10^{14}$	$1\ 10^7$
Sn-125	$1\ 10^2$	$1\ 10^5$	$1\ 10^{10}$	$1\ 10^6$
Sn-126+	$1\ 10^1$	$1\ 10^5$	$1\ 10^{10}$	$1\ 10^6$
Sn-127	$1\ 10^1$	$1\ 10^6$	$1\ 10^{12}$	$1\ 10^7$
Sn-128	$1\ 10^1$	$1\ 10^6$	$1\ 10^{13}$	$1\ 10^7$

1 Radionuclide name, symbol, isotope	2 Concentration for notification. Regulation 6 and Schedule 1 (Bq/g)	3 Quantity for notification. Regulation 6 and Schedule 1 (Bq)	4 Quantity for notification of occurrences. Regulation 30(1) (Bq)	5 Quantity for notification of occurrences. Regulation 30(3) (Bq)
Antimony				
Sb-115	$1\ 10^1$	$1\ 10^6$	$1\ 10^{13}$	$1\ 10^7$
Sb-116	$1\ 10^1$	$1\ 10^6$	$1\ 10^{13}$	$1\ 10^7$
Sb-116m	$1\ 10^1$	$1\ 10^5$	$1\ 10^{12}$	$1\ 10^6$
Sb-117	$1\ 10^2$	$1\ 10^7$	$1\ 10^{13}$	$1\ 10^8$
Sb-118m	$1\ 10^1$	$1\ 10^6$	$1\ 10^{12}$	$1\ 10^7$
Sb-119	$1\ 10^3$	$1\ 10^7$	$1\ 10^{12}$	$1\ 10^8$
Sb-120 (5.76 days)	$1\ 10^1$	$1\ 10^6$	$1\ 10^{10}$	$1\ 10^7$
Sb-120 (15.89 min)	$1\ 10^2$	$1\ 10^6$	$1\ 10^{14}$	$1\ 10^7$
Sb-122	$1\ 10^2$	$1\ 10^4$	$1\ 10^{11}$	$1\ 10^5$
Sb-124	$1\ 10^1$	$1\ 10^6$	$1\ 10^{10}$	$1\ 10^7$
Sb-124m	$1\ 10^2$	$1\ 10^6$	$1\ 10^{14}$	$1\ 10^7$
Sb-125	$1\ 10^2$	$1\ 10^6$	$1\ 10^{10}$	$1\ 10^7$
Sb-126	$1\ 10^1$	$1\ 10^5$	$1\ 10^{10}$	$1\ 10^6$
Sb-126m	$1\ 10^1$	$1\ 10^5$	$1\ 10^{13}$	$1\ 10^6$
Sb-127	$1\ 10^1$	$1\ 10^6$	$1\ 10^{11}$	$1\ 10^7$
Sb-128 (9.01 hours)	$1\ 10^1$	$1\ 10^5$	$1\ 10^{11}$	$1\ 10^6$
Sb-128 (10.4 min)	$1\ 10^1$	$1\ 10^5$	$1\ 10^{13}$	$1\ 10^6$
Sb-129	$1\ 10^1$	$1\ 10^6$	$1\ 10^{12}$	$1\ 10^7$
Sb-130	$1\ 10^1$	$1\ 10^5$	$1\ 10^{13}$	$1\ 10^6$
Sb-131	$1\ 10^1$	$1\ 10^6$	$1\ 10^{13}$	$1\ 10^7$
Tellurium				
Te-116	$1\ 10^2$	$1\ 10^7$	$1\ 10^{13}$	$1\ 10^8$
Te-121	$1\ 10^1$	$1\ 10^6$	$1\ 10^{11}$	$1\ 10^7$
Te-121m	$1\ 10^2$	$1\ 10^6$	$1\ 10^{10}$	$1\ 10^7$
Te-123	$1\ 10^3$	$1\ 10^6$	$1\ 10^{10}$	$1\ 10^7$
Te-123m	$1\ 10^2$	$1\ 10^7$	$1\ 10^{10}$	$1\ 10^8$
Te-125m	$1\ 10^3$	$1\ 10^7$	$1\ 10^{10}$	$1\ 10^8$
Te-127	$1\ 10^3$	$1\ 10^6$	$1\ 10^{12}$	$1\ 10^7$
Te-127m	$1\ 10^3$	$1\ 10^7$	$1\ 10^{10}$	$1\ 10^8$
Te-129	$1\ 10^2$	$1\ 10^6$	$1\ 10^{14}$	$1\ 10^7$
Te-129m	$1\ 10^3$	$1\ 10^6$	$1\ 10^{10}$	$1\ 10^7$
Te-131	$1\ 10^2$	$1\ 10^5$	$1\ 10^{14}$	$1\ 10^6$
Te-131m	$1\ 10^1$	$1\ 10^6$	$1\ 10^{11}$	$1\ 10^7$
Te-132	$1\ 10^2$	$1\ 10^7$	$1\ 10^{11}$	$1\ 10^8$
Te-133	$1\ 10^1$	$1\ 10^5$	$1\ 10^{14}$	$1\ 10^6$
Te-133m	$1\ 10^1$	$1\ 10^5$	$1\ 10^{13}$	$1\ 10^6$
Te-134	$1\ 10^1$	$1\ 10^6$	$1\ 10^{13}$	$1\ 10^7$
Iodine				
I-120	$1\ 10^1$	$1\ 10^5$	$1\ 10^{12}$	$1\ 10^6$
I-120m	$1\ 10^1$	$1\ 10^5$	$1\ 10^{12}$	$1\ 10^6$
I-121	$1\ 10^2$	$1\ 10^6$	$1\ 10^{13}$	$1\ 10^7$
I-123	$1\ 10^2$	$1\ 10^7$	$1\ 10^{12}$	$1\ 10^8$
I-124	$1\ 10^1$	$1\ 10^6$	$1\ 10^{10}$	$1\ 10^7$
I-125	$1\ 10^3$	$1\ 10^6$	$1\ 10^{10}$	$1\ 10^7$
I-126	$1\ 10^2$	$1\ 10^6$	$1\ 10^{10}$	$1\ 10^7$
I-128	$1\ 10^2$	$1\ 10^5$	$1\ 10^{14}$	$1\ 10^6$
I-129	$1\ 10^2$	$1\ 10^5$	$1\ 10^9$	$1\ 10^6$
I-130	$1\ 10^1$	$1\ 10^6$	$1\ 10^{11}$	$1\ 10^7$
I-131	$1\ 10^2$	$1\ 10^6$	$1\ 10^{10}$	$1\ 10^7$
I-132	$1\ 10^1$	$1\ 10^5$	$1\ 10^{12}$	$1\ 10^6$
I-132m	$1\ 10^2$	$1\ 10^6$	$1\ 10^{13}$	$1\ 10^7$
I-133	$1\ 10^1$	$1\ 10^6$	$1\ 10^{11}$	$1\ 10^7$
I-134	$1\ 10^1$	$1\ 10^5$	$1\ 10^{13}$	$1\ 10^6$
I-135	$1\ 10^1$	$1\ 10^6$	$1\ 10^{12}$	$1\ 10^7$

1 Radionuclide name, symbol, isotope	2 Concentration for notification. Regulation 6 and Schedule 1 (Bq/g)	3 Quantity for notification. Regulation 6 and Schedule 1 (Bq)	4 Quantity for notification of occurrences. Regulation 30(1) (Bq)	5 Quantity for notification of occurrences. Regulation 30(3) (Bq)
Xenon				
Xe-120	$1\ 10^2$	$1\ 10^9$	$1\ 10^{10}$	
Xe-121	$1\ 10^2$	$1\ 10^9$	$1\ 10^9$	
Xe-122+	$1\ 10^2$	$1\ 10^9$	$1\ 10^{11}$	
Xe-123	$1\ 10^2$	$1\ 10^9$	$1\ 10^9$	
Xe-125	$1\ 10^3$	$1\ 10^9$	$1\ 10^{10}$	
Xe-127	$1\ 10^3$	$1\ 10^5$	$1\ 10^{10}$	
Xe-129m	$1\ 10^3$	$1\ 10^4$	$1\ 10^{11}$	
Xe-131m	$1\ 10^4$	$1\ 10^4$	$1\ 10^{11}$	
Xe-133	$1\ 10^3$	$1\ 10^4$	$1\ 10^{11}$	
Xe-133m	$1\ 10^3$	$1\ 10^4$	$1\ 10^{11}$	
Xe-135	$1\ 10^3$	$1\ 10^{10}$	$1\ 10^{10}$	
Xe-135m	$1\ 10^2$	$1\ 10^9$	$1\ 10^{10}$	
Xe-138	$1\ 10^2$	$1\ 10^9$	$1\ 10^9$	
Caesium				
Cs-125	$1\ 10^1$	$1\ 10^4$	$1\ 10^{13}$	$1\ 10^5$
Cs-127	$1\ 10^2$	$1\ 10^5$	$1\ 10^{12}$	$1\ 10^6$
Cs-129	$1\ 10^2$	$1\ 10^5$	$1\ 10^{12}$	$1\ 10^6$
Cs-130	$1\ 10^2$	$1\ 10^6$	$1\ 10^{14}$	$1\ 10^7$
Cs-131	$1\ 10^3$	$1\ 10^6$	$1\ 10^{12}$	$1\ 10^7$
Cs-132	$1\ 10^1$	$1\ 10^5$	$1\ 10^{11}$	$1\ 10^6$
Cs-134	$1\ 10^1$	$1\ 10^4$	$1\ 10^{10}$	$1\ 10^5$
Cs-134m	$1\ 10^3$	$1\ 10^5$	$1\ 10^{14}$	$1\ 10^6$
Cs-135	$1\ 10^4$	$1\ 10^7$	$1\ 10^{11}$	$1\ 10^8$
Cs-135m	$1\ 10^1$	$1\ 10^6$	$1\ 10^{13}$	$1\ 10^7$
Cs-136	$1\ 10^1$	$1\ 10^5$	$1\ 10^{10}$	$1\ 10^6$
Cs-137+	$1\ 10^1$	$1\ 10^4$	$1\ 10^{10}$	$1\ 10^5$
Cs-138	$1\ 10^1$	$1\ 10^4$	$1\ 10^{13}$	$1\ 10^5$
Barium				
Ba-126	$1\ 10^2$	$1\ 10^7$	$1\ 10^{13}$	$1\ 10^8$
Ba-128	$1\ 10^2$	$1\ 10^7$	$1\ 10^{11}$	$1\ 10^8$
Ba-131	$1\ 10^2$	$1\ 10^6$	$1\ 10^{11}$	$1\ 10^7$
Ba-131m	$1\ 10^2$	$1\ 10^7$	$1\ 10^{15}$	$1\ 10^8$
Ba-133	$1\ 10^2$	$1\ 10^6$	$1\ 10^{11}$	$1\ 10^7$
Ba-133m	$1\ 10^2$	$1\ 10^6$	$1\ 10^{12}$	$1\ 10^7$
Ba-135m	$1\ 10^2$	$1\ 10^6$	$1\ 10^{12}$	$1\ 10^7$
Ba-137m	$1\ 10^1$	$1\ 10^6$	$1\ 10^{15}$	$1\ 10^7$
Ba-139	$1\ 10^2$	$1\ 10^5$	$1\ 10^{13}$	$1\ 10^6$
Ba-140+	$1\ 10^1$	$1\ 10^5$	$1\ 10^{11}$	$1\ 10^6$
Ba-141	$1\ 10^1$	$1\ 10^5$	$1\ 10^{13}$	$1\ 10^6$
Ba-142	$1\ 10^1$	$1\ 10^6$	$1\ 10^{14}$	$1\ 10^7$
Lanthanum				
La-131	$1\ 10^1$	$1\ 10^6$	$1\ 10^{13}$	$1\ 10^7$
La-132	$1\ 10^1$	$1\ 10^6$	$1\ 10^{12}$	$1\ 10^7$
La-135	$1\ 10^3$	$1\ 10^7$	$1\ 10^{13}$	$1\ 10^8$
La-137	$1\ 10^3$	$1\ 10^7$	$1\ 10^{10}$	$1\ 10^8$
La-138	$1\ 10^1$	$1\ 10^6$	$1\ 10^9$	$1\ 10^7$
La-140	$1\ 10^1$	$1\ 10^5$	$1\ 10^{11}$	$1\ 10^6$
La-141	$1\ 10^2$	$1\ 10^5$	$1\ 10^{13}$	$1\ 10^6$
La-142	$1\ 10^1$	$1\ 10^5$	$1\ 10^{12}$	$1\ 10^6$
La-143	$1\ 10^2$	$1\ 10^5$	$1\ 10^{14}$	$1\ 10^6$

1 Radionuclide name, symbol, isotope	2 Concentration for notification. Regulation 6 and Schedule 1 (Bq/g)	3 Quantity for notification. Regulation 6 and Schedule 1 (Bq)	4 Quantity for notification of occurrences. Regulation 30(1) (Bq)	5 Quantity for notification of occurrences. Regulation 30(3) (Bq)
Cerium				
Ce-134	$1\ 10^3$	$1\ 10^7$	$1\ 10^{11}$	$1\ 10^8$
Ce-135	$1\ 10^1$	$1\ 10^6$	$1\ 10^{11}$	$1\ 10^7$
Ce-137	$1\ 10^3$	$1\ 10^7$	$1\ 10^{13}$	$1\ 10^8$
Ce-137m	$1\ 10^3$	$1\ 10^6$	$1\ 10^{11}$	$1\ 10^7$
Ce-139	$1\ 10^2$	$1\ 10^6$	$1\ 10^{11}$	$1\ 10^7$
Ce-141	$1\ 10^2$	$1\ 10^7$	$1\ 10^{10}$	$1\ 10^8$
Ce-143	$1\ 10^2$	$1\ 10^6$	$1\ 10^{11}$	$1\ 10^7$
Ce-144+	$1\ 10^2$	$1\ 10^5$	$1\ 10^9$	$1\ 10^6$
Praseody-mium				
Pr-136	$1\ 10^1$	$1\ 10^5$	$1\ 10^{13}$	$1\ 10^6$
Pr-137	$1\ 10^2$	$1\ 10^6$	$1\ 10^{13}$	$1\ 10^7$
Pr-138m	$1\ 10^1$	$1\ 10^6$	$1\ 10^{12}$	$1\ 10^7$
Pr-139	$1\ 10^2$	$1\ 10^7$	$1\ 10^{13}$	$1\ 10^8$
Pr-142	$1\ 10^2$	$1\ 10^5$	$1\ 10^{12}$	$1\ 10^6$
Pr-142m	$1\ 10^7$	$1\ 10^9$	$1\ 10^{15}$	$1\ 10^{10}$
Pr-143	$1\ 10^4$	$1\ 10^6$	$1\ 10^{11}$	$1\ 10^7$
Pr-144	$1\ 10^2$	$1\ 10^5$	$1\ 10^{14}$	$1\ 10^6$
Pr-145	$1\ 10^3$	$1\ 10^5$	$1\ 10^{12}$	$1\ 10^6$
Pr-147	$1\ 10^1$	$1\ 10^5$	$1\ 10^{14}$	$1\ 10^6$
Neodymium				
Nd-136	$1\ 10^2$	$1\ 10^6$	$1\ 10^{13}$	$1\ 10^7$
Nd-138	$1\ 10^3$	$1\ 10^7$	$1\ 10^{12}$	$1\ 10^8$
Nd-139	$1\ 10^2$	$1\ 10^6$	$1\ 10^{14}$	$1\ 10^7$
Nd-139m	$1\ 10^1$	$1\ 10^6$	$1\ 10^{12}$	$1\ 10^7$
Nd-141	$1\ 10^2$	$1\ 10^7$	$1\ 10^{14}$	$1\ 10^8$
Nd-147	$1\ 10^2$	$1\ 10^6$	$1\ 10^{11}$	$1\ 10^7$
Nd-149	$1\ 10^2$	$1\ 10^6$	$1\ 10^{13}$	$1\ 10^7$
Nd-151	$1\ 10^1$	$1\ 10^5$	$1\ 10^{14}$	$1\ 10^6$
Promethium				
Pm-141	$1\ 10^1$	$1\ 10^5$	$1\ 10^{13}$	$1\ 10^6$
Pm-143	$1\ 10^2$	$1\ 10^6$	$1\ 10^{11}$	$1\ 10^7$
Pm-144	$1\ 10^1$	$1\ 10^6$	$1\ 10^{10}$	$1\ 10^7$
Pm-145	$1\ 10^3$	$1\ 10^7$	$1\ 10^{10}$	$1\ 10^8$
Pm-146	$1\ 10^1$	$1\ 10^6$	$1\ 10^{10}$	$1\ 10^7$
Pm-147	$1\ 10^4$	$1\ 10^7$	$1\ 10^{10}$	$1\ 10^8$
Pm-148	$1\ 10^1$	$1\ 10^5$	$1\ 10^{11}$	$1\ 10^6$
Pm-148m+	$1\ 10^1$	$1\ 10^6$	$1\ 10^{10}$	$1\ 10^7$
Pm-149	$1\ 10^3$	$1\ 10^6$	$1\ 10^{11}$	$1\ 10^7$
Pm-150	$1\ 10^1$	$1\ 10^5$	$1\ 10^{12}$	$1\ 10^6$
Pm-151	$1\ 10^2$	$1\ 10^6$	$1\ 10^{11}$	$1\ 10^7$
Samarium				
Sm-141	$1\ 10^1$	$1\ 10^5$	$1\ 10^{13}$	$1\ 10^6$
Sm-141m	$1\ 10^1$	$1\ 10^6$	$1\ 10^{13}$	$1\ 10^7$
Sm-142	$1\ 10^2$	$1\ 10^7$	$1\ 10^{13}$	$1\ 10^8$
Sm-145	$1\ 10^2$	$1\ 10^7$	$1\ 10^{11}$	$1\ 10^8$
Sm-146	$1\ 10^1$	$1\ 10^5$	$1\ 10^7$	$1\ 10^6$
Sm-147	$1\ 10^1$	$1\ 10^4$	$1\ 10^7$	$1\ 10^5$
Sm-151	$1\ 10^4$	$1\ 10^8$	$1\ 10^{10}$	$1\ 10^9$
Sm-153	$1\ 10^2$	$1\ 10^6$	$1\ 10^{11}$	$1\ 10^7$
Sm-155	$1\ 10^2$	$1\ 10^6$	$1\ 10^{14}$	$1\ 10^7$
Sm-156	$1\ 10^2$	$1\ 10^6$	$1\ 10^{12}$	$1\ 10^7$

1 Radionuclide name, symbol, isotope	2 Concentration for notification. Regulation 6 and Schedule 1 (Bq/g)	3 Quantity for notification. Regulation 6 and Schedule 1 (Bq)	4 Quantity for notification of occurrences. Regulation 30(1) (Bq)	5 Quantity for notification of occurrences. Regulation 30(3) (Bq)
Europium				
Eu-145	$1\ 10^1$	$1\ 10^6$	$1\ 10^{11}$	$1\ 10^7$
Eu-146	$1\ 10^1$	$1\ 10^6$	$1\ 10^{11}$	$1\ 10^7$
Eu-147	$1\ 10^2$	$1\ 10^6$	$1\ 10^{11}$	$1\ 10^7$
Eu-148	$1\ 10^1$	$1\ 10^6$	$1\ 10^{10}$	$1\ 10^7$
Eu-149	$1\ 10^2$	$1\ 10^7$	$1\ 10^{11}$	$1\ 10^8$
Eu-150 (34.2 yrs)	$1\ 10^1$	$1\ 10^6$	$1\ 10^9$	$1\ 10^7$
Eu-150 (12.6 hrs)	$1\ 10^3$	$1\ 10^6$	$1\ 10^{12}$	$1\ 10^7$
Eu-152	$1\ 10^1$	$1\ 10^6$	$1\ 10^9$	$1\ 10^7$
Eu-152m	$1\ 10^2$	$1\ 10^6$	$1\ 10^{12}$	$1\ 10^7$
Eu-154	$1\ 10^1$	$1\ 10^6$	$1\ 10^9$	$1\ 10^7$
Eu-155	$1\ 10^2$	$1\ 10^7$	$1\ 10^{10}$	$1\ 10^8$
Eu-156	$1\ 10^1$	$1\ 10^6$	$1\ 10^{10}$	$1\ 10^7$
Eu-157	$1\ 10^2$	$1\ 10^6$	$1\ 10^{12}$	$1\ 10^7$
Eu-158	$1\ 10^1$	$1\ 10^5$	$1\ 10^{13}$	$1\ 10^6$
Gadolinium				
Gd-145	$1\ 10^1$	$1\ 10^5$	$1\ 10^{13}$	$1\ 10^6$
Gd-146+	$1\ 10^1$	$1\ 10^6$	$1\ 10^{10}$	$1\ 10^7$
Gd-147	$1\ 10^1$	$1\ 10^6$	$1\ 10^{11}$	$1\ 10^7$
Gd-148	$1\ 10^1$	$1\ 10^4$	$1\ 10^6$	$1\ 10^5$
Gd-149	$1\ 10^2$	$1\ 10^6$	$1\ 10^{11}$	$1\ 10^7$
Gd-151	$1\ 10^2$	$1\ 10^7$	$1\ 10^{11}$	$1\ 10^8$
Gd-152	$1\ 10^1$	$1\ 10^4$	$1\ 10^6$	$1\ 10^5$
Gd-153	$1\ 10^2$	$1\ 10^7$	$1\ 10^{10}$	$1\ 10^8$
Gd-159	$1\ 10^3$	$1\ 10^6$	$1\ 10^{12}$	$1\ 10^7$
Terbium				
Tb-147	$1\ 10^1$	$1\ 10^6$	$1\ 10^{12}$	$1\ 10^7$
Tb-149	$1\ 10^1$	$1\ 10^6$	$1\ 10^{11}$	$1\ 10^7$
Tb-150	$1\ 10^1$	$1\ 10^6$	$1\ 10^{12}$	$1\ 10^7$
Tb-151	$1\ 10^1$	$1\ 10^6$	$1\ 10^{12}$	$1\ 10^7$
Tb-153	$1\ 10^2$	$1\ 10^7$	$1\ 10^{12}$	$1\ 10^8$
Tb-154	$1\ 10^1$	$1\ 10^6$	$1\ 10^{11}$	$1\ 10^7$
Tb-155	$1\ 10^2$	$1\ 10^7$	$1\ 10^{11}$	$1\ 10^8$
Tb-156	$1\ 10^1$	$1\ 10^6$	$1\ 10^{11}$	$1\ 10^7$
Tb-156m(24.4 hrs)	$1\ 10^3$	$1\ 10^7$	$1\ 10^{12}$	$1\ 10^8$
Tb-156m (5 hrs)	$1\ 10^4$	$1\ 10^7$	$1\ 10^{13}$	$1\ 10^8$
Tb-157	$1\ 10^4$	$1\ 10^7$	$1\ 10^{11}$	$1\ 10^8$
Tb-158	$1\ 10^1$	$1\ 10^6$	$1\ 10^9$	$1\ 10^7$
Tb-160	$1\ 10^1$	$1\ 10^6$	$1\ 10^{10}$	$1\ 10^7$
Tb-161	$1\ 10^3$	$1\ 10^6$	$1\ 10^{11}$	$1\ 10^7$
Dysprosium				
Dy-155	$1\ 10^1$	$1\ 10^6$	$1\ 10^{12}$	$1\ 10^7$
Dy-157	$1\ 10^2$	$1\ 10^6$	$1\ 10^{12}$	$1\ 10^7$
Dy-159	$1\ 10^3$	$1\ 10^7$	$1\ 10^{11}$	$1\ 10^8$
Dy-165	$1\ 10^3$	$1\ 10^6$	$1\ 10^{13}$	$1\ 10^7$
Dy-166	$1\ 10^3$	$1\ 10^6$	$1\ 10^{11}$	$1\ 10^7$

1 Radionuclide name, symbol, isotope	2 Concentration for notification. Regulation 6 and Schedule 1 (Bq/g)	3 Quantity for notification. Regulation 6 and Schedule 1 (Bq)	4 Quantity for notification of occurrences. Regulation 30(1) (Bq)	5 Quantity for notification of occurrences. Regulation 30(3) (Bq)
Holmium				
Ho-155	$1\ 10^2$	$1\ 10^6$	$1\ 10^{13}$	$1\ 10^7$
Ho-157	$1\ 10^2$	$1\ 10^6$	$1\ 10^{14}$	$1\ 10^7$
Ho-159	$1\ 10^2$	$1\ 10^6$	$1\ 10^{14}$	$1\ 10^7$
Ho-161	$1\ 10^2$	$1\ 10^7$	$1\ 10^{14}$	$1\ 10^8$
Ho-162	$1\ 10^2$	$1\ 10^7$	$1\ 10^{14}$	$1\ 10^8$
Ho-162m	$1\ 10^1$	$1\ 10^6$	$1\ 10^{13}$	$1\ 10^7$
Ho-164	$1\ 10^3$	$1\ 10^6$	$1\ 10^{14}$	$1\ 10^7$
Ho-164m	$1\ 10^3$	$1\ 10^7$	$1\ 10^{14}$	$1\ 10^8$
Ho-166	$1\ 10^3$	$1\ 10^5$	$1\ 10^{11}$	$1\ 10^6$
Ho-166m	$1\ 10^1$	$1\ 10^6$	$1\ 10^9$	$1\ 10^7$
Ho-167	$1\ 10^2$	$1\ 10^6$	$1\ 10^{13}$	$1\ 10^7$
Erbium				
Er-161	$1\ 10^1$	$1\ 10^6$	$1\ 10^{12}$	$1\ 10^7$
Er-165	$1\ 10^3$	$1\ 10^7$	$1\ 10^{13}$	$1\ 10^8$
Er-169	$1\ 10^4$	$1\ 10^7$	$1\ 10^{11}$	$1\ 10^8$
Er-171	$1\ 10^2$	$1\ 10^6$	$1\ 10^{12}$	$1\ 10^7$
Er-172	$1\ 10^2$	$1\ 10^6$	$1\ 10^{11}$	$1\ 10^7$
Thulium				
Tm-162	$1\ 10^1$	$1\ 10^6$	$1\ 10^{13}$	$1\ 10^7$
Tm-166	$1\ 10^1$	$1\ 10^6$	$1\ 10^{12}$	$1\ 10^7$
Tm-167	$1\ 10^2$	$1\ 10^6$	$1\ 10^{11}$	$1\ 10^7$
Tm-170	$1\ 10^3$	$1\ 10^6$	$1\ 10^{10}$	$1\ 10^7$
Tm-171	$1\ 10^4$	$1\ 10^8$	$1\ 10^{11}$	$1\ 10^9$
Tm-172	$1\ 10^2$	$1\ 10^6$	$1\ 10^{11}$	$1\ 10^7$
Tm-173	$1\ 10^2$	$1\ 10^6$	$1\ 10^{12}$	$1\ 10^7$
Tm-175	$1\ 10^1$	$1\ 10^6$	$1\ 10^{13}$	$1\ 10^7$
Ytterbium				
Yb-162	$1\ 10^2$	$1\ 10^7$	$1\ 10^{14}$	$1\ 10^8$
Yb-166	$1\ 10^2$	$1\ 10^7$	$1\ 10^{11}$	$1\ 10^8$
Yb-167	$1\ 10^2$	$1\ 10^6$	$1\ 10^{14}$	$1\ 10^7$
Yb-169	$1\ 10^2$	$1\ 10^7$	$1\ 10^{10}$	$1\ 10^8$
Yb-175	$1\ 10^3$	$1\ 10^7$	$1\ 10^{11}$	$1\ 10^8$
Yb-177	$1\ 10^2$	$1\ 10^6$	$1\ 10^{13}$	$1\ 10^7$
Yb-178	$1\ 10^3$	$1\ 10^6$	$1\ 10^{13}$	$1\ 10^7$
Lutetium				
Lu-169	$1\ 10^1$	$1\ 10^6$	$1\ 10^{11}$	$1\ 10^7$
Lu-170	$1\ 10^1$	$1\ 10^6$	$1\ 10^{11}$	$1\ 10^7$
Lu-171	$1\ 10^1$	$1\ 10^6$	$1\ 10^{11}$	$1\ 10^7$
Lu-172	$1\ 10^1$	$1\ 10^6$	$1\ 10^{10}$	$1\ 10^7$
Lu-173	$1\ 10^2$	$1\ 10^7$	$1\ 10^{11}$	$1\ 10^8$
Lu-174	$1\ 10^2$	$1\ 10^7$	$1\ 10^{10}$	$1\ 10^8$
Lu-174m	$1\ 10^2$	$1\ 10^7$	$1\ 10^{10}$	$1\ 10^8$
Lu-176	$1\ 10^2$	$1\ 10^6$	$1\ 10^9$	$1\ 10^7$
Lu-176m	$1\ 10^3$	$1\ 10^6$	$1\ 10^{13}$	$1\ 10^7$
Lu-177	$1\ 10^3$	$1\ 10^7$	$1\ 10^{11}$	$1\ 10^8$
Lu-177m	$1\ 10^1$	$1\ 10^6$	$1\ 10^{10}$	$1\ 10^7$
Lu-178	$1\ 10^2$	$1\ 10^5$	$1\ 10^{14}$	$1\ 10^6$
Lu-178m	$1\ 10^1$	$1\ 10^5$	$1\ 10^{13}$	$1\ 10^6$
Lu-179	$1\ 10^3$	$1\ 10^6$	$1\ 10^{13}$	$1\ 10^7$

1 Radionuclide name, symbol, isotope	2 Concentration for notification. Regulation 6 and Schedule 1 (Bq/g)	3 Quantity for notification. Regulation 6 and Schedule 1 (Bq)	4 Quantity for notification of occurrences. Regulation 30(1) (Bq)	5 Quantity for notification of occurrences. Regulation 30(3) (Bq)
Hafnium				
Hf-170	$1\ 10^2$	$1\ 10^6$	$1\ 10^{12}$	$1\ 10^7$
Hf-172+	$1\ 10^1$	$1\ 10^6$	$1\ 10^9$	$1\ 10^7$
Hf-173	$1\ 10^2$	$1\ 10^6$	$1\ 10^{12}$	$1\ 10^7$
Hf-175	$1\ 10^2$	$1\ 10^6$	$1\ 10^{11}$	$1\ 10^7$
Hf-177m	$1\ 10^1$	$1\ 10^5$	$1\ 10^{13}$	$1\ 10^6$
Hf-178m	$1\ 10^1$	$1\ 10^6$	$1\ 10^8$	$1\ 10^7$
Hf-179m	$1\ 10^1$	$1\ 10^6$	$1\ 10^{10}$	$1\ 10^7$
Hf-180m	$1\ 10^1$	$1\ 10^6$	$1\ 10^{12}$	$1\ 10^7$
Hf-181	$1\ 10^1$	$1\ 10^6$	$1\ 10^{10}$	$1\ 10^7$
Hf-182	$1\ 10^2$	$1\ 10^6$	$1\ 10^8$	$1\ 10^7$
Hf-182m	$1\ 10^1$	$1\ 10^6$	$1\ 10^{13}$	$1\ 10^7$
Hf-183	$1\ 10^1$	$1\ 10^6$	$1\ 10^{13}$	$1\ 10^7$
Hf-184	$1\ 10^2$	$1\ 10^6$	$1\ 10^{12}$	$1\ 10^7$
Tantalum				
Ta-172	$1\ 10^1$	$1\ 10^6$	$1\ 10^{13}$	$1\ 10^7$
Ta-173	$1\ 10^1$	$1\ 10^6$	$1\ 10^{12}$	$1\ 10^7$
Ta-174	$1\ 10^1$	$1\ 10^6$	$1\ 10^{13}$	$1\ 10^7$
Ta-175	$1\ 10^1$	$1\ 10^6$	$1\ 10^{10}$	$1\ 10^7$
Ta-176	$1\ 10^1$	$1\ 10^6$	$1\ 10^{12}$	$1\ 10^7$
Ta-177	$1\ 10^2$	$1\ 10^7$	$1\ 10^{12}$	$1\ 10^8$
Ta-178	$1\ 10^1$	$1\ 10^6$	$1\ 10^{13}$	$1\ 10^7$
Ta-179	$1\ 10^3$	$1\ 10^7$	$1\ 10^{11}$	$1\ 10^8$
Ta-180	$1\ 10^1$	$1\ 10^6$	$1\ 10^{10}$	$1\ 10^7$
Ta-180m	$1\ 10^3$	$1\ 10^7$	$1\ 10^{13}$	$1\ 10^8$
Ta-182	$1\ 10^1$	$1\ 10^4$	$1\ 10^{10}$	$1\ 10^5$
Ta-182m	$1\ 10^2$	$1\ 10^6$	$1\ 10^{14}$	$1\ 10^7$
Ta-183	$1\ 10^2$	$1\ 10^6$	$1\ 10^{11}$	$1\ 10^7$
Ta-184	$1\ 10^1$	$1\ 10^6$	$1\ 10^{12}$	$1\ 10^7$
Ta-185	$1\ 10^2$	$1\ 10^5$	$1\ 10^{13}$	$1\ 10^6$
Ta-186	$1\ 10^1$	$1\ 10^5$	$1\ 10^{13}$	$1\ 10^6$
Tungsten				
W-176	$1\ 10^2$	$1\ 10^6$	$1\ 10^{13}$	$1\ 10^7$
W-177	$1\ 10^1$	$1\ 10^6$	$1\ 10^{13}$	$1\ 10^7$
W-178+	$1\ 10^1$	$1\ 10^6$	$1\ 10^{12}$	$1\ 10^7$
W-179	$1\ 10^2$	$1\ 10^7$	$1\ 10^{14}$	$1\ 10^8$
W-181	$1\ 10^3$	$1\ 10^7$	$1\ 10^{12}$	$1\ 10^8$
W-185	$1\ 10^4$	$1\ 10^7$	$1\ 10^{11}$	$1\ 10^8$
W-187	$1\ 10^2$	$1\ 10^6$	$1\ 10^{12}$	$1\ 10^7$
W-188+	$1\ 10^2$	$1\ 10^5$	$1\ 10^{11}$	$1\ 10^6$
Rhenium				
Re-177	$1\ 10^1$	$1\ 10^6$	$1\ 10^{14}$	$1\ 10^7$
Re-178	$1\ 10^1$	$1\ 10^6$	$1\ 10^{13}$	$1\ 10^7$
Re-181	$1\ 10^1$	$1\ 10^6$	$1\ 10^{11}$	$1\ 10^7$
Re-182 (64 hours)	$1\ 10^1$	$1\ 10^6$	$1\ 10^{11}$	$1\ 10^7$
Re-182 (12.7 hrs)	$1\ 10^1$	$1\ 10^6$	$1\ 10^{12}$	$1\ 10^7$
Re-184	$1\ 10^1$	$1\ 10^6$	$1\ 10^{10}$	$1\ 10^7$
Re-184m	$1\ 10^2$	$1\ 10^6$	$1\ 10^{10}$	$1\ 10^7$
Re-186	$1\ 10^3$	$1\ 10^6$	$1\ 10^{11}$	$1\ 10^7$
Re-186m	$1\ 10^3$	$1\ 10^7$	$1\ 10^{10}$	$1\ 10^8$
Re-187	$1\ 10^6$	$1\ 10^9$	$1\ 10^{13}$	$1\ 10^{10}$
Re-188	$1\ 10^2$	$1\ 10^5$	$1\ 10^{12}$	$1\ 10^6$
Re-188m	$1\ 10^2$	$1\ 10^7$	$1\ 10^{14}$	$1\ 10^8$
Re-189+	$1\ 10^2$	$1\ 10^6$	$1\ 10^{12}$	$1\ 10^7$

1 Radionuclide name, symbol, isotope	2 Concentration for notification. Regulation 6 and Schedule 1 (Bq/g)	3 Quantity for notification. Regulation 6 and Schedule 1 (Bq)	4 Quantity for notification of occurrences. Regulation 30(1) (Bq)	5 Quantity for notification of occurrences. Regulation 30(3) (Bq)
Osmium				
Os-180	$1\ 10^2$	$1\ 10^7$	$1\ 10^{14}$	$1\ 10^8$
Os-181	$1\ 10^1$	$1\ 10^6$	$1\ 10^{13}$	$1\ 10^7$
Os-182	$1\ 10^2$	$1\ 10^6$	$1\ 10^{11}$	$1\ 10^7$
Os-185	$1\ 10^1$	$1\ 10^6$	$1\ 10^{11}$	$1\ 10^7$
Os-189m	$1\ 10^4$	$1\ 10^7$	$1\ 10^{14}$	$1\ 10^8$
Os-191	$1\ 10^2$	$1\ 10^7$	$1\ 10^{11}$	$1\ 10^8$
Os-191m	$1\ 10^3$	$1\ 10^7$	$1\ 10^{12}$	$1\ 10^8$
Os-193	$1\ 10^2$	$1\ 10^6$	$1\ 10^{11}$	$1\ 10^7$
Os-194+	$1\ 10^2$	$1\ 10^5$	$1\ 10^9$	$1\ 10^6$
Iridium				
Ir-182	$1\ 10^1$	$1\ 10^5$	$1\ 10^{13}$	$1\ 10^6$
Ir-184	$1\ 10^1$	$1\ 10^6$	$1\ 10^{12}$	$1\ 10^7$
Ir-185	$1\ 10^1$	$1\ 10^6$	$1\ 10^{12}$	$1\ 10^7$
Ir-186 (15.8 hrs)	$1\ 10^1$	$1\ 10^6$	$1\ 10^{11}$	$1\ 10^7$
Ir-186 (1.75 hrs)	$1\ 10^1$	$1\ 10^6$	$1\ 10^{13}$	$1\ 10^7$
Ir-187	$1\ 10^2$	$1\ 10^6$	$1\ 10^{12}$	$1\ 10^7$
Ir-188	$1\ 10^1$	$1\ 10^6$	$1\ 10^{11}$	$1\ 10^7$
Ir-189+	$1\ 10^2$	$1\ 10^7$	$1\ 10^{11}$	$1\ 10^8$
Ir-190	$1\ 10^1$	$1\ 10^6$	$1\ 10^{10}$	$1\ 10^7$
Ir-190m (3.1 hrs)	$1\ 10^1$	$1\ 10^6$	$1\ 10^{13}$	$1\ 10^7$
Ir-190m (1.2 hrs)	$1\ 10^4$	$1\ 10^7$	$1\ 10^{15}$	$1\ 10^8$
Ir-192	$1\ 10^1$	$1\ 10^4$	$1\ 10^{10}$	$1\ 10^5$
Ir-192m	$1\ 10^2$	$1\ 10^7$	$1\ 10^{10}$	$1\ 10^8$
Ir-193m	$1\ 10^4$	$1\ 10^7$	$1\ 10^{11}$	$1\ 10^8$
Ir-194	$1\ 10^2$	$1\ 10^5$	$1\ 10^{11}$	$1\ 10^6$
Ir-194m	$1\ 10^1$	$1\ 10^6$	$1\ 10^{10}$	$1\ 10^7$
Ir-195	$1\ 10^2$	$1\ 10^6$	$1\ 10^{13}$	$1\ 10^7$
Ir-195m	$1\ 10^2$	$1\ 10^6$	$1\ 10^{12}$	$1\ 10^7$
Platinum				
Pt-186	$1\ 10^1$	$1\ 10^6$	$1\ 10^{13}$	$1\ 10^7$
Pt-188+	$1\ 10^1$	$1\ 10^6$	$1\ 10^{11}$	$1\ 10^7$
Pt-189	$1\ 10^2$	$1\ 10^6$	$1\ 10^{12}$	$1\ 10^7$
Pt-191	$1\ 10^2$	$1\ 10^6$	$1\ 10^{11}$	$1\ 10^7$
Pt-193	$1\ 10^4$	$1\ 10^7$	$1\ 10^{12}$	$1\ 10^8$
Pt-193m	$1\ 10^3$	$1\ 10^7$	$1\ 10^{12}$	$1\ 10^8$
Pt-195m	$1\ 10^2$	$1\ 10^6$	$1\ 10^{11}$	$1\ 10^7$
Pt-197	$1\ 10^3$	$1\ 10^6$	$1\ 10^{12}$	$1\ 10^7$
Pt-197m	$1\ 10^2$	$1\ 10^6$	$1\ 10^{14}$	$1\ 10^7$
Pt-199	$1\ 10^2$	$1\ 10^6$	$1\ 10^{14}$	$1\ 10^7$
Pt-200	$1\ 10^2$	$1\ 10^6$	$1\ 10^{12}$	$1\ 10^7$
Gold				
Au-193	$1\ 10^2$	$1\ 10^7$	$1\ 10^{12}$	$1\ 10^8$
Au-194	$1\ 10^1$	$1\ 10^6$	$1\ 10^{11}$	$1\ 10^7$
Au-195	$1\ 10^2$	$1\ 10^7$	$1\ 10^{11}$	$1\ 10^8$
Au-198	$1\ 10^2$	$1\ 10^6$	$1\ 10^{11}$	$1\ 10^7$
Au-198m	$1\ 10^1$	$1\ 10^6$	$1\ 10^{11}$	$1\ 10^7$
Au-199	$1\ 10^2$	$1\ 10^6$	$1\ 10^{11}$	$1\ 10^7$
Au-200	$1\ 10^2$	$1\ 10^5$	$1\ 10^{13}$	$1\ 10^6$
Au-200m	$1\ 10^1$	$1\ 10^6$	$1\ 10^{11}$	$1\ 10^7$
Au-201	$1\ 10^2$	$1\ 10^6$	$1\ 10^{14}$	$1\ 10^7$

1 Radionuclide name, symbol, isotope	2 Concentration for notification. Regulation 6 and Schedule 1 (Bq/g)	3 Quantity for notification. Regulation 6 and Schedule 1 (Bq)	4 Quantity for notification of occurrences. Regulation 30(1) (Bq)	5 Quantity for notification of occurrences. Regulation 30(3) (Bq)
Mercury				
Hg-193	$1\ 10^2$	$1\ 10^6$	$1\ 10^{13}$	$1\ 10^7$
Hg-193m	$1\ 10^1$	$1\ 10^6$	$1\ 10^{12}$	$1\ 10^7$
Hg-194+	$1\ 10^1$	$1\ 10^6$	$1\ 10^{10}$	$1\ 10^7$
Hg-195	$1\ 10^2$	$1\ 10^6$	$1\ 10^{12}$	$1\ 10^7$
Hg-195m+(organic)	$1\ 10^2$	$1\ 10^6$	$1\ 10^{12}$	$1\ 10^7$
Hg-195m+(inorganic)	$1\ 10^2$	$1\ 10^6$	$1\ 10^{11}$	$1\ 10^7$
Hg-197	$1\ 10^2$	$1\ 10^7$	$1\ 10^{12}$	$1\ 10^8$
Hg-197m (organic)	$1\ 10^2$	$1\ 10^6$	$1\ 10^{12}$	$1\ 10^7$
Hg-197m (inorganic)	$1\ 10^2$	$1\ 10^6$	$1\ 10^{11}$	$1\ 10^7$
Hg-199m	$1\ 10^2$	$1\ 10^6$	$1\ 10^{14}$	$1\ 10^7$
Hg-203	$1\ 10^2$	$1\ 10^5$	$1\ 10^{11}$	$1\ 10^6$
Thallium				
Tl-194	$1\ 10^1$	$1\ 10^6$	$1\ 10^{13}$	$1\ 10^7$
Tl-194m	$1\ 10^1$	$1\ 10^6$	$1\ 10^{13}$	$1\ 10^7$
Tl-195	$1\ 10^1$	$1\ 10^6$	$1\ 10^{13}$	$1\ 10^7$
Tl-197	$1\ 10^2$	$1\ 10^6$	$1\ 10^{13}$	$1\ 10^7$
Tl-198	$1\ 10^1$	$1\ 10^6$	$1\ 10^{12}$	$1\ 10^7$
Tl-198m	$1\ 10^1$	$1\ 10^6$	$1\ 10^{13}$	$1\ 10^7$
Tl-199	$1\ 10^2$	$1\ 10^6$	$1\ 10^{13}$	$1\ 10^7$
Tl-200	$1\ 10^1$	$1\ 10^6$	$1\ 10^{11}$	$1\ 10^7$
Tl-201	$1\ 10^2$	$1\ 10^6$	$1\ 10^{12}$	$1\ 10^7$
Tl-202	$1\ 10^2$	$1\ 10^6$	$1\ 10^{11}$	$1\ 10^7$
Tl-204	$1\ 10^4$	$1\ 10^4$	$1\ 10^{11}$	$1\ 10^5$
Lead				
Pb-195m	$1\ 10^1$	$1\ 10^6$	$1\ 10^{13}$	$1\ 10^7$
Pb-198	$1\ 10^2$	$1\ 10^6$	$1\ 10^{13}$	$1\ 10^7$
Pb-199	$1\ 10^1$	$1\ 10^6$	$1\ 10^{13}$	$1\ 10^7$
Pb-200	$1\ 10^2$	$1\ 10^6$	$1\ 10^{12}$	$1\ 10^7$
Pb-201	$1\ 10^1$	$1\ 10^6$	$1\ 10^{12}$	$1\ 10^7$
Pb-202	$1\ 10^3$	$1\ 10^6$	$1\ 10^{10}$	$1\ 10^7$
Pb-202m	$1\ 10^1$	$1\ 10^6$	$1\ 10^{12}$	$1\ 10^7$
Pb-203	$1\ 10^2$	$1\ 10^6$	$1\ 10^{12}$	$1\ 10^7$
Pb-205	$1\ 10^4$	$1\ 10^7$	$1\ 10^{11}$	$1\ 10^8$
Pb-209	$1\ 10^5$	$1\ 10^6$	$1\ 10^{14}$	$1\ 10^7$
Pb-210+	$1\ 10^1$	$1\ 10^4$	$1\ 10^8$	$1\ 10^5$
Pb-211	$1\ 10^2$	$1\ 10^6$	$1\ 10^{12}$	$1\ 10^7$
Pb-212+	$1\ 10^1$	$1\ 10^5$	$1\ 10^{10}$	$1\ 10^6$
Pb-214	$1\ 10^2$	$1\ 10^6$	$1\ 10^{12}$	$1\ 10^7$
Bismuth				
Bi-200	$1\ 10^1$	$1\ 10^6$	$1\ 10^{13}$	$1\ 10^7$
Bi-201	$1\ 10^1$	$1\ 10^6$	$1\ 10^{12}$	$1\ 10^7$
Bi-202	$1\ 10^1$	$1\ 10^6$	$1\ 10^{12}$	$1\ 10^7$
Bi-203	$1\ 10^1$	$1\ 10^6$	$1\ 10^{11}$	$1\ 10^7$
Bi-205	$1\ 10^1$	$1\ 10^6$	$1\ 10^{11}$	$1\ 10^7$
Bi-206	$1\ 10^1$	$1\ 10^5$	$1\ 10^{10}$	$1\ 10^6$
Bi-207	$1\ 10^1$	$1\ 10^6$	$1\ 10^{10}$	$1\ 10^7$
Bi-210	$1\ 10^3$	$1\ 10^6$	$1\ 10^9$	$1\ 10^7$
Bi-210m+	$1\ 10^1$	$1\ 10^5$	$1\ 10^8$	$1\ 10^6$
Bi-212+	$1\ 10^1$	$1\ 10^5$	$1\ 10^{11}$	$1\ 10^6$
Bi-213	$1\ 10^2$	$1\ 10^6$	$1\ 10^{11}$	$1\ 10^7$
Bi-214	$1\ 10^1$	$1\ 10^5$	$1\ 10^{12}$	$1\ 10^6$

1 Radionuclide name, symbol, isotope	2 Concentration for notification. Regulation 6 and Schedule 1 (Bq/g)	3 Quantity for notification. Regulation 6 and Schedule 1 (Bq)	4 Quantity for notification of occurrences. Regulation 30(1) (Bq)	5 Quantity for notification of occurrences. Regulation 30(3) (Bq)
Polonium				
Po-203	$1\ 10^1$	$1\ 10^6$	$1\ 10^{13}$	$1\ 10^7$
Po-205	$1\ 10^1$	$1\ 10^6$	$1\ 10^{12}$	$1\ 10^7$
Po-206	$1\ 10^1$	$1\ 10^6$	$1\ 10^{11}$	$1\ 10^7$
Po-207	$1\ 10^1$	$1\ 10^6$	$1\ 10^{12}$	$1\ 10^7$
Po-208	$1\ 10^1$	$1\ 10^4$	$1\ 10^7$	$1\ 10^5$
Po-209	$1\ 10^1$	$1\ 10^4$	$1\ 10^7$	$1\ 10^5$
Po-210	$1\ 10^1$	$1\ 10^4$	$1\ 10^7$	$1\ 10^5$
Astatine				
At-207	$1\ 10^1$	$1\ 10^6$	$1\ 10^{12}$	$1\ 10^7$
At-211	$1\ 10^3$	$1\ 10^7$	$1\ 10^{10}$	$1\ 10^8$
Francium				
Fr-222	$1\ 10^3$	$1\ 10^5$	$1\ 10^{12}$	$1\ 10^6$
Fr-223	$1\ 10^2$	$1\ 10^6$	$1\ 10^{13}$	$1\ 10^7$
Radon				
Rn-220+	$1\ 10^4$	$1\ 10^7$	$1\ 10^8$	$1\ 10^8$
Rn-222+	$1\ 10^1$	$1\ 10^8$	$1\ 10^9$	$1\ 10^9$
Radium				
Ra-223+	$1\ 10^2$	$1\ 10^5$	$1\ 10^7$	$1\ 10^6$
Ra-224+	$1\ 10^1$	$1\ 10^5$	$1\ 10^8$	$1\ 10^6$
Ra-225	$1\ 10^2$	$1\ 10^5$	$1\ 10^7$	$1\ 10^6$
Ra-226+	$1\ 10^1$	$1\ 10^4$	$1\ 10^7$	$1\ 10^5$
Ra-227	$1\ 10^2$	$1\ 10^6$	$1\ 10^{13}$	$1\ 10^7$
Ra-228+	$1\ 10^1$	$1\ 10^5$	$1\ 10^8$	$1\ 10^6$
Actinium				
Ac-224	$1\ 10^2$	$1\ 10^6$	$1\ 10^{10}$	$1\ 10^7$
Ac-225+	$1\ 10^1$	$1\ 10^4$	$1\ 10^7$	$1\ 10^5$
Ac-226	$1\ 10^2$	$1\ 10^5$	$1\ 10^8$	$1\ 10^6$
Ac-227+	$1\ 10^{-1}$	$1\ 10^3$	$1\ 10^5$	$1\ 10^4$
Ac-228	$1\ 10^1$	$1\ 10^6$	$1\ 10^{10}$	$1\ 10^7$
Thorium				
Th-226+	$1\ 10^3$	$1\ 10^7$	$1\ 10^{11}$	$1\ 10^8$
Th-227	$1\ 10^1$	$1\ 10^4$	$1\ 10^7$	$1\ 10^5$
Th-228+	$1\ 10^0$	$1\ 10^4$	$1\ 10^6$	$1\ 10^5$
Th-229+	$1\ 10^0$	$1\ 10^3$	$1\ 10^6$	$1\ 10^4$
Th-230	$1\ 10^0$	$1\ 10^4$	$1\ 10^6$	$1\ 10^5$
Th-231	$1\ 10^3$	$1\ 10^7$	$1\ 10^{12}$	$1\ 10^8$
Th-232	$1\ 10^1$	$1\ 10^4$	$1\ 10^6$	$1\ 10^5$
Th-232sec	$1\ 10^0$	$1\ 10^3$	$1\ 10^6$	$1\ 10^4$
Th-234+	$1\ 10^3$	$1\ 10^5$	$1\ 10^{10}$	$1\ 10^6$
Protactinium				
Pa-227	$1\ 10^3$	$1\ 10^6$	$1\ 10^{11}$	$1\ 10^7$
Pa-228	$1\ 10^1$	$1\ 10^6$	$1\ 10^{10}$	$1\ 10^7$
Pa-230	$1\ 10^1$	$1\ 10^6$	$1\ 10^8$	$1\ 10^7$
Pa-231	$1\ 10^0$	$1\ 10^3$	$1\ 10^6$	$1\ 10^4$
Pa-232	$1\ 10^1$	$1\ 10^6$	$1\ 10^{10}$	$1\ 10^7$
Pa-233	$1\ 10^2$	$1\ 10^7$	$1\ 10^{10}$	$1\ 10^8$
Pa-234	$1\ 10^1$	$1\ 10^6$	$1\ 10^{12}$	$1\ 10^7$

1 Radionuclide name, symbol, isotope	2 Concentration for notification. Regulation 6 and Schedule 1 (Bq/g)	3 Quantity for notification. Regulation 6 and Schedule 1 (Bq)	4 Quantity for notification of occurrences. Regulation 30(1) (Bq)	5 Quantity for notification of occurrences. Regulation 30(3) (Bq)
Uranium				
U-230+	$1\ 10^1$	$1\ 10^5$	$1\ 10^7$	$1\ 10^6$
U-231	$1\ 10^2$	$1\ 10^7$	$1\ 10^{11}$	$1\ 10^8$
U-232+	$1\ 10^0$	$1\ 10^3$	$1\ 10^6$	$1\ 10^4$
U-233	$1\ 10^1$	$1\ 10^4$	$1\ 10^7$	$1\ 10^5$
U-234	$1\ 10^1$	$1\ 10^4$	$1\ 10^7$	$1\ 10^5$
U-235+	$1\ 10^1$	$1\ 10^4$	$1\ 10^7$	$1\ 10^5$
U-236	$1\ 10^1$	$1\ 10^4$	$1\ 10^7$	$1\ 10^5$
U-237	$1\ 10^2$	$1\ 10^6$	$1\ 10^{11}$	$1\ 10^7$
U-238+	$1\ 10^1$	$1\ 10^4$	$1\ 10^7$	$1\ 10^5$
U-238 sec	$1\ 10^0$	$1\ 10^3$	$1\ 10^6$	$1\ 10^4$
U-239	$1\ 10^2$	$1\ 10^6$	$1\ 10^{14}$	$1\ 10^7$
U-240	$1\ 10^3$	$1\ 10^7$	$1\ 10^{12}$	$1\ 10^8$
U-240+	$1\ 10^1$	$1\ 10^6$	$1\ 10^{11}$	$1\ 10^7$
Neptunium				
Np-232	$1\ 10^1$	$1\ 10^6$	$1\ 10^{13}$	$1\ 10^7$
Np-233	$1\ 10^2$	$1\ 10^7$	$1\ 10^{14}$	$1\ 10^8$
Np-234	$1\ 10^1$	$1\ 10^6$	$1\ 10^{11}$	$1\ 10^7$
Np-235	$1\ 10^3$	$1\ 10^7$	$1\ 10^{11}$	$1\ 10^8$
Np-236(1.15 10^5 yrs)	$1\ 10^2$	$1\ 10^5$	$1\ 10^8$	$1\ 10^6$
Np-236(22.5 hrs)	$1\ 10^3$	$1\ 10^7$	$1\ 10^{11}$	$1\ 10^8$
Np-237+	$1\ 10^0$	$1\ 10^3$	$1\ 10^7$	$1\ 10^4$
Np-238	$1\ 10^2$	$1\ 10^6$	$1\ 10^{11}$	$1\ 10^7$
Np-239	$1\ 10^2$	$1\ 10^7$	$1\ 10^{11}$	$1\ 10^8$
Np-240	$1\ 10^1$	$1\ 10^6$	$1\ 10^{13}$	$1\ 10^7$
Plutonium				
Pu-234	$1\ 10^2$	$1\ 10^7$	$1\ 10^{10}$	$1\ 10^8$
Pu-235	$1\ 10^2$	$1\ 10^7$	$1\ 10^{14}$	$1\ 10^8$
Pu-236	$1\ 10^1$	$1\ 10^4$	$1\ 10^7$	$1\ 10^5$
Pu-237	$1\ 10^3$	$1\ 10^7$	$1\ 10^{11}$	$1\ 10^8$
Pu-238	$1\ 10^0$	$1\ 10^4$	$1\ 10^6$	$1\ 10^5$
Pu-239	$1\ 10^0$	$1\ 10^4$	$1\ 10^6$	$1\ 10^5$
Pu-240	$1\ 10^0$	$1\ 10^3$	$1\ 10^6$	$1\ 10^4$
Pu-241	$1\ 10^2$	$1\ 10^5$	$1\ 10^8$	$1\ 10^6$
Pu-242	$1\ 10^0$	$1\ 10^4$	$1\ 10^6$	$1\ 10^5$
Pu-243	$1\ 10^3$	$1\ 10^7$	$1\ 10^{13}$	$1\ 10^8$
Pu-244	$1\ 10^0$	$1\ 10^4$	$1\ 10^6$	$1\ 10^5$
Pu-245	$1\ 10^2$	$1\ 10^6$	$1\ 10^{12}$	$1\ 10^7$
Pu-246	$1\ 10^2$	$1\ 10^6$	$1\ 10^{10}$	$1\ 10^7$
Americium				
Am-237	$1\ 10^2$	$1\ 10^6$	$1\ 10^{13}$	$1\ 10^7$
Am-238	$1\ 10^1$	$1\ 10^6$	$1\ 10^{13}$	$1\ 10^7$
Am-239	$1\ 10^2$	$1\ 10^6$	$1\ 10^{12}$	$1\ 10^7$
Am-240	$1\ 10^1$	$1\ 10^6$	$1\ 10^{11}$	$1\ 10^7$
Am-241	$1\ 10^0$	$1\ 10^4$	$1\ 10^6$	$1\ 10^5$
Am-242	$1\ 10^3$	$1\ 10^6$	$1\ 10^{10}$	$1\ 10^7$
Am-242m+	$1\ 10^0$	$1\ 10^4$	$1\ 10^6$	$1\ 10^5$
Am-243+	$1\ 10^0$	$1\ 10^3$	$1\ 10^6$	$1\ 10^4$
Am-244	$1\ 10^1$	$1\ 10^6$	$1\ 10^{11}$	$1\ 10^7$
Am-244m	$1\ 10^4$	$1\ 10^7$	$1\ 10^{14}$	$1\ 10^8$
Am-245	$1\ 10^3$	$1\ 10^6$	$1\ 10^{13}$	$1\ 10^7$
Am-246	$1\ 10^1$	$1\ 10^5$	$1\ 10^{13}$	$1\ 10^6$
Am-246m	$1\ 10^1$	$1\ 10^6$	$1\ 10^{13}$	$1\ 10^7$

1 Radionuclide name, symbol, isotope	2 Concentration for notification. Regulation 6 and Schedule 1 (Bq/g)	3 Quantity for notification. Regulation 6 and Schedule 1 (Bq)	4 Quantity for notification of occurrences. Regulation 30(1) (Bq)	5 Quantity for notification of occurrences. Regulation 30(3) (Bq)
Curium				
Cm-238	$1\ 10^2$	$1\ 10^7$	$1\ 10^{12}$	$1\ 10^8$
Cm-240	$1\ 10^2$	$1\ 10^5$	$1\ 10^7$	$1\ 10^6$
Cm-241	$1\ 10^2$	$1\ 10^6$	$1\ 10^9$	$1\ 10^7$
Cm-242	$1\ 10^2$	$1\ 10^5$	$1\ 10^7$	$1\ 10^6$
Cm-243	$1\ 10^0$	$1\ 10^4$	$1\ 10^7$	$1\ 10^5$
Cm-244	$1\ 10^1$	$1\ 10^4$	$1\ 10^7$	$1\ 10^5$
Cm-245	$1\ 10^0$	$1\ 10^3$	$1\ 10^6$	$1\ 10^4$
Cm-246	$1\ 10^0$	$1\ 10^3$	$1\ 10^6$	$1\ 10^4$
Cm-247	$1\ 10^0$	$1\ 10^4$	$1\ 10^6$	$1\ 10^5$
Cm-248	$1\ 10^0$	$1\ 10^3$	$1\ 10^6$	$1\ 10^4$
Cm-249	$1\ 10^3$	$1\ 10^6$	$1\ 10^{14}$	$1\ 10^7$
Cm-250	$1\ 10^{-1}$	$1\ 10^3$	$1\ 10^5$	$1\ 10^4$
Berkelium				
Bk-245	$1\ 10^2$	$1\ 10^6$	$1\ 10^{11}$	$1\ 10^7$
Bk-246	$1\ 10^1$	$1\ 10^6$	$1\ 10^{11}$	$1\ 10^7$
Bk-247	$1\ 10^0$	$1\ 10^4$	$1\ 10^6$	$1\ 10^5$
Bk-249	$1\ 10^3$	$1\ 10^6$	$1\ 10^9$	$1\ 10^7$
Bk-250	$1\ 10^1$	$1\ 10^6$	$1\ 10^{12}$	$1\ 10^7$
Californium				
Cf-244	$1\ 10^4$	$1\ 10^7$	$1\ 10^{12}$	$1\ 10^8$
Cf-246	$1\ 10^3$	$1\ 10^6$	$1\ 10^9$	$1\ 10^7$
Cf-248	$1\ 10^1$	$1\ 10^4$	$1\ 10^7$	$1\ 10^5$
Cf-249	$1\ 10^0$	$1\ 10^3$	$1\ 10^6$	$1\ 10^4$
Cf-250	$1\ 10^1$	$1\ 10^4$	$1\ 10^6$	$1\ 10^5$
Cf-251	$1\ 10^0$	$1\ 10^3$	$1\ 10^6$	$1\ 10^4$
Cf-252	$1\ 10^1$	$1\ 10^4$	$1\ 10^7$	$1\ 10^5$
Cf-253	$1\ 10^2$	$1\ 10^5$	$1\ 10^8$	$1\ 10^6$
Cf-254	$1\ 10^0$	$1\ 10^3$	$1\ 10^7$	$1\ 10^4$
Einsteinium				
Es-250	$1\ 10^2$	$1\ 10^6$	$1\ 10^{13}$	$1\ 10^7$
Es-251	$1\ 10^2$	$1\ 10^7$	$1\ 10^{11}$	$1\ 10^8$
Es-253	$1\ 10^2$	$1\ 10^5$	$1\ 10^8$	$1\ 10^6$
Es-254	$1\ 10^1$	$1\ 10^4$	$1\ 10^7$	$1\ 10^5$
Es-254m	$1\ 10^2$	$1\ 10^6$	$1\ 10^9$	$1\ 10^7$
Fermium				
Fm-252	$1\ 10^3$	$1\ 10^6$	$1\ 10^9$	$1\ 10^7$
Fm-253	$1\ 10^2$	$1\ 10^6$	$1\ 10^9$	$1\ 10^7$
Fm-254	$1\ 10^4$	$1\ 10^7$	$1\ 10^{10}$	$1\ 10^8$
Fm-255	$1\ 10^3$	$1\ 10^6$	$1\ 10^9$	$1\ 10^7$
Fm-257	$1\ 10^1$	$1\ 10^5$	$1\ 10^7$	$1\ 10^6$
Mendelevium				
Md-257	$1\ 10^2$	$1\ 10^7$	$1\ 10^{11}$	$1\ 10^8$
Md-258	$1\ 10^2$	$1\ 10^5$	$1\ 10^7$	$1\ 10^6$
Other radio-nuclides not listed above (see note 1)	$1\ 10^{-1}$	$1\ 10^3$	$1\ 10^5$	$1\ 10^4$

Note 1
In the case of radionuclides not specified elsewhere in this Part, the quantities specified in this entry are to be used unless the Executive has approved some other quantity for that radionuclide.

Note 2
Nuclides carrying the suffix '+' or 'sec' in the above table represent parent nuclides in equilibrium with their correspondent daughter nuclides as listed in the following Table. In this case the concentrations and quantities given in the above Table refer to the parent nuclide alone, but already take account of the daughter nuclide(s) present.

List of nuclides in secular equilibrium as referred to in note 2 of this Schedule.

Parent nuclide	Daughter nuclides
Mg-28+	Al-28
Ti-44+	Sc-44
Fe-60+	Co-60m
Ge-68+	Ga-68
Sr-82+	Rb-82
Rb-83+	Kr-83m
Y-87+	Sr-87m
Sr-90+	Y-90
Zr-93+	Nb-93m
Zr-97+	Nb-97
Tc-95m+	Tc-95
Ru-106+	Rh-106
Ag-108m+	Ag-108
Sn-121m+	Sn-121
Sn-126+	Sb-126m
Xe-122+	I-122
Cs-137+	Ba-137m
Ba-140+	La-140
Ce-144+	Pr-144
Pm-148m+	Pm-148
Gd-146+	Eu-146
Hf-172+	Lu-172
W-178+	Ta-178
W-188+	Re-188
Re-189+	Os-189m
Os-194+	Ir-194
Ir-189+	Os-189m
Pt-188+	Ir-188
Hg-194+	Au-194
Hg-195m+	Hg-195
Pb-210+	Bi-210, Po-210
Bi-210m+	Tl-206
Pb-212+	Bi-212, Tl-208, Po-212
Bi-212+	Tl-208, Po-212
Rn-220+	Po-216
Rn-222+	Po-218, Pb-214, Bi-214, Po-214
Ra-223+	Rn-219, Po-215, Pb-211, Bi-211, Tl-207
Ra-224+	Rn-220, Po-216, Pb-212, Bi-212, Tl-208, Po-212
Ra-226+	Rn-222, Po-218, Pb-214, Bi-214, Po-214, Pb-210, Bi-210, Po-210
Ra-228+	Ac-228
Ac-225+	Fr-221, At-217, Bi-213, Po-213, Tl-209, Pb-209
Ac-227+	Fr-223
Th-226+	Ra-222, Rn-218, Po-214
Th-228+	Ra-224, Rn-220, Po-216, Pb-212, Bi-212, Tl-208, Po-212
Th-229+	Ra-225, Ac-225, Fr-221, At-217, Bi-213, Po-213, Pb-209
Th-232sec	Ra-228, Ac-228, Th-228, Ra-224, Rn-220, Po-216, Pb-212, Bi-212, Tl-208, Po-212
Th-234+	Pa-234m
U-230+	Th-226, Ra-222, Rn-218, Po-214

U-232+	Th-228, Ra-224, Rn-220, Po-216, Pb-212, Bi-212, Tl-208, Po-212
U-235+	Th-231
U-238+	Th-234, Pa-234m
U-238sec	Th-234, Pa-234m, U-234, Th-230, Ra-226, Rn-222, Po-218, Pb-214, Bi-214, Po-214, Pb-210, Bi-210, Po-210
U-240+	Np-240
Np-237+	Pa-233
Am-242m+	Am-242
Am-243+	Np-239

Part II Quantity ratios for more than one radionuclide

1   For the purpose of Regulation 2(4), the quantity ratio for more than one radionuclide is the sum of the quotients of the quantity of a radionuclide present Q_p divided by the quantity of that radionuclide specified in the appropriate column of Part I of this Schedule Q_{lim}, namely -

$$\sum \frac{Q_p}{Q_{lim}}$$

2   In any case where the isotopic composition of a radioactive substance is not known or is only partially known, the quantity ratio for that substance shall be calculated by using the values specified in the appropriate column in Part 1 for 'other radionuclides not listed above' for any radionuclide that has not been identified or where the quantity of a radionuclide is uncertain, unless the employer can show that the use of some other value is appropriate in the circumstances of a particular case, when he may use that value.

Schedule 9

Modifications

Regulation 41(1)

The Employment Act 1989

1   In Schedule 1 to the Employment Act 1989,[a] in place of "Parts IV and V of the Ionising Radiations Regulations 1985" there shall be substituted "Paragraphs 5 and 11 of Schedule 4 to the Ionising Radiations Regulations 1999 [S.I. 1999/3232]".

The Employment Rights Act 1996

2   In section 64(3) of the Employment Rights Act 1996,[b] in place of "regulation 16 of the Ionising Radiations Regulations 1985" there shall be substituted "regulation 24 of the Ionising Radiations Regulations 1999 [S.I. 1999/3232]".

The Personal Protective Equipment at Work Regulations 1992

3   In regulation 3 of the Personal Protective Equipment at Work Regulations 1992,[c] in place of "the Ionising Radiations Regulations 1985" there shall be substituted "the Ionising Radiations Regulations 1999 [S.I. 1999/3232]".

(a) 1989 c. 38.   (b) 1996 c. 18.   (c) SI 1992/2966.

The Notification of New Substances Regulations 1993

4 In sub-paragraph (2)(e) of regulation 3 of the Notification of New Substances Regulations 1993,[a] in place of "the Ionising Radiations Regulations 1985" there shall be substituted "the Ionising Radiations Regulations 1999 [S.I. 1999/3232]".

The Chemicals (Hazard Information and Packaging for Supply) Regulations 1994

5 In sub-paragraph (1)(a) of regulation 3 of the Chemicals (Hazard Information and Packaging for Supply) Regulations 1994,[b] in place of the "the Ionising Radiations Regulations 1985" there shall be substituted "the Ionising Radiations Regulations 1999 [S.I. 1999/3232]".

The Reporting of Injuries, Diseases and Dangerous Occurrences Regulations 1995

6 Paragraph 8(2) of Schedule 2 to the Reporting of Injuries, Diseases and Dangerous Occurrences Regulations 1995[c] shall be deleted and the following substituted -

"In this paragraph, "radiation generator" means any electrical equipment emitting ionising radiation and containing components operating at a potential difference of more than 5 kV."

The Packaging, Labelling and Carriage of Radioactive Material by Rail Regulations 1996

7 In paragraph 1(s) of Schedule 14 to the Packaging, Labelling and Carriage of Radioactive Material by Rail Regulations 1996,[d] after "regulation 27 of the Ionising Radiations Regulations 1985" there shall be added "or regulation 12 of the Ionising Radiations Regulations 1999 [S.I. 1999/3232]".

The Health and Safety (Enforcing Authority) Regulations 1998

8 The Health and Safety (Enforcing Authority) Regulations 1998[e] shall be modified as follows -

(a) in the definition of "ionising radiation" in regulation 2(1), in place of "the Ionising Radiations Regulations 1985" there shall be substituted "the Ionising Radiations Regulations 1999 [S.I. 1999/3232]";

(b) in sub-paragraph (d) of paragraph 4 of Schedule 2, in place of "Schedule 3 0f the Ionising Radiations Regulations 1985" there shall be substituted "Schedule 1 of the Ionising Radiations Regulations 1999 [S.I. 1999/3232]",

(c) in paragraph 5 of Schedule 2, in place of "the Ionising Radiations Regulations 1985" there shall be substituted "the Ionising Radiations Regulations 1999 [S.I. 1999/3232]".

(a) SI 1993/3050.
(b) SI 1994/3247.
(c) SI 1995/3162.
(d) SI 1996/2090.
(e) SI 1998/494.

The Health and Safety (Fees) Regulations 1999

9   The Health and Safety (Fees) Regulations 1999[a] shall be modified as follows -

(a)   in sub-paragraph (1)(c) of regulation 3, in place of "the Ionising Radiations Regulations 1985" there shall be substituted "the Ionising Radiations Regulations 1999 [S.I. 1999/3232]";

(b)   the chapeau to regulation 9 shall be deleted and the following substituted -

"Fees for application for approval or reassessment of approval of dosimetry services and for type approval of apparatus under the Ionising Radiations Regulations 1999";

(c)   in regulation 9(2), the words "a radiation generator or an apparatus containing a radioactive substance" shall be deleted and substituted by the following -

"apparatus pursuant to sub-paragraphs 1(c)(i) and 1(d)(i) of Schedule 1 to the Ionising Radiations Regulations 1999 [S.I. 1999/3232]";

(d)   the title of Schedule 8 shall be deleted and the following substituted -

"FEES FOR APPLICATIONS FOR APPROVAL OR REASSESSMENT OF APPROVAL OF DOSIMETRY SERVICES AND FOR TYPE APPROVAL OF APPARATUS UNDER THE IONISING RADIATIONS REGULATIONS 1999";

(e)   in the first entry of column 1 of Schedule 8, in place of "regulation 15 of the Ionising Radiations Regulations 1985", there shall be substituted "regulation 35 of the Ionising Radiations Regulations 1999 [S.I. 1999/3232]";

(f)   the final entry of column 1 of Schedule 8 shall be deleted and the following substituted -

"Type approval of apparatus under sub-paragraph 1(c)(i) or 1(d)(i) of Schedule 1 to the Ionising Radiations Regulations 1999 [S.I. 1999/3232] (which excepts such type approved apparatus from the notification requirements of regulation 6 of those Regulations)."

(a) SI 1999/645.

Estimating effective dose and equivalent dose from external radiation

Effective dose

Effective dose from external radiation, is generally defined by the relationship:

$$E_{ext} = \sum_T w_T H_T = \sum_T w_T \sum_R w_R D_{T,R}$$

This represents the sum of the weighted equivalent doses in all the tissues and organs of the body from external radiation, where:

- $D_{T,R}$ is the absorbed dose averaged over the tissue or organ T, due to radiation R

  H_T is the equivalent dose

- w_R is the radiation weighting factor, and

- w_T is the tissue weighting factor for tissue or organ T.

The total effective dose, E, includes the 50-year committed effective dose from internal radiation, E_{int}, from inhaled or ingested radionuclides usually based on radionuclide-specific dose per unit intake values (Annex III of Directive 96/29/Euratom gives default values).

Equivalent dose

Equivalent dose (H_T) is the absorbed dose, in tissue or organ T weighted for the type and quality of radiation R. It is given by:

$$H_{T,R} = w_R D_{T,R}$$

where:

- $D_{T,R}$ is the absorbed dose averaged over the tissue or organ T, due to radiation R,

- w_R is the radiation weighting factor.

When the radiation field is composed of types and energies with different values of w_R, the total equivalent dose, H_T is given by:

$$H_T = \sum_R w_R D_{T,R}$$

Estimating effective dose

Where the effective dose from external radiation is estimated from personal monitoring, the operational dose quantity *personal dose equivalent* $H_P(d)$ is normally used to demonstrate compliance with dose limits.

Personal dose equivalent is the dose in soft tissues at an appropriate depth, d in mm, below a specified point in the body and is given in sieverts.

Where doses are to be estimated from area monitoring results, the relevant operational quantities are:

- ambient dose equivalent $H^{\star}(d)$; and

- directional dose equivalent $H'(d,\Omega)$.

In these cases, d is the depth in mm under the surface of the International Commission on Radiation Units and Measurements (ICRU) sphere. For strongly penetrating radiation a depth of 10 mm is appropriate and for weakly penetrating radiation a depth of 0.07 mm for the skin and 3 mm for the eye is recommended. Ω is the angle of incidence.

Ambient dose equivalent is the dose equivalent at a point in a radiation field that would be produced by the corresponding expanded and aligned field in the ICRU sphere at a depth, d, on the radius opposing the direction of the aligned field and is given in sieverts.

Directional dose equivalent is the dose equivalent at a point in a radiation field that would be produced by the corresponding expanded field, in the ICRU sphere at a depth, d, on a radius in a specified direction, Ω, and is given in sieverts.

The *ICRU sphere* is a body introduced by the ICRU to approximate the human body as regards energy absorption from ionising radiation; it consists of a 30 cm diameter tissue equivalent sphere with a density of 1g cm^{-3} and a mass composition of 76.2% oxygen, 11.1% carbon, 10.1% hydrogen and 2.6% nitrogen.

An *expanded field* is a field derived from the actual field, where the fluence and its directional and energy distributions have the same values throughout the volume of interest as in the actual field at the point of reference.

An *expanded and aligned field* is a radiation field in which the fluence and its directional and energy distribution are the same as in the expanded field but the fluence is unidirectional.

The *fluence*, Φ is the quotient of dN by da, where dN is the number of particles which enter a sphere of cross-sectional area da.

The *quality factor*, Q, is a function of linear energy transfer (L) in water and is used to weight the absorbed dose at a point in such a way as to take into account the quality of a radiation in a specified tissue or organ.

Unrestricted linear energy transfer (L) is the quotient of dE by dl, where dE is the mean energy lost by a particle of energy E in traversing a distance dl in water.

The relationship between the quality factor Q(L) and the unrestricted energy transfer L is as follows.

Unrestricted linear energy transfer, L, in water (keV μ m^{-1})	Q(L)
< 10	1
10-100	0.32L-2.2
> 100	$300\sqrt{L}$

Tissue weighting factors

Appropriate values of tissue weighting factor (w_T) to be used to weight the equivalent dose in a tissue or organ (T), where necessary.

Tissue or organ	Tissue weighting factor
Gonads	0.2
Bone marrow (red)	0.12
Colon*	0.12
Lung	0.12
Stomach	0.12
Bladder	0.05
Breast	0.05
Liver	0.05
Oesophagus	0.05
Thyroid	0.05
Skin	0.01
Bone surface	0.01
Remainder	0.05(†)(††)

*   Dose to the colon is taken to be the mass weighted average dose to the upper and lower large intestines.

†   For the purposes of calculation, the remainder is composed of the following additional tissues and organs: adrenals, brain, small intestine, kidneys, muscle, pancreas, spleen, thymus, uterus and extrathoracic airways.

††   The equivalent dose to the remainder tissues is normally calculated as the mass-weighted mean dose to the ten organs and tissues listed above. In the exceptional case in which the most highly irradiated remainder tissue or organ receives the highest equivalent dose of all organs, a weighting factor of 0.025 (half of remainder) is applied to that tissue or organ and 0.025 to the mass weighted equivalent dose in the rest of the remainder tissues and organs.

Radiation weighting factors

Where required for a direct estimate of E, values of radiation weighting factor, W_R, depend on the type and quality of the external radiation field.

Type and energy range	Radiation weighting factor, W_R
Photons, all energies	1
Electrons and muons, all energies	1
Neutrons, energy, < 10 keV	5
10 keV to 100 keV	10
> 100 keV to 2 MeV	20
> 2 MeV to 20 MeV	10
> 20 MeV	5
Protons, other than recoil protons, energy > 2MeV	5
Alpha particles, fission fragments, heavy nuclei	20

In calculations involving neutrons it may be preferable to use:

$$W_R = 5 + 17e^{-(\ln(2E))^2/6}$$

where E is the neutron energy.

HSE may authorise the use of equivalent methods for estimating E.

Explanation of terms

Appointed doctor	See definition in regulation 2(1)
Approved dosimetry services (ADS)	See definition in regulation 2(1) and paragraphs 382-384
Classified person	See definition in regulation 2(1) and paragraphs 367-370
Comforter and carer	See definition in regulation 2(1) and paragraphs 127 and 128
Committed effective dose	See *effective dose* and paragraphs 182 and 186
Contingency plan	See paragraphs 206-213
Controlled area	See definition in regulation 2(1) and guidance to regulation 16(1)
Designated area	A controlled area or supervised area - see guidance to regulation 16
Dose	Any dose quantity or sum of dose quantities mentioned in Schedule 4
Dose assessment	The dose assessment made and recorded by an ADS in relation to a classified person in accordance with regulation 21 (see also *Requirements for the approval of dosimetry services* Parts 1 & 2)[41, 42]
Dose assessment period	See paragraph 389
Dose constraint	See definition in regulation 2(1) and paragraph 119
Dose limit	See definition in regulation 2(1)
Dose rate	See definition in regulation 2(1)
Dose record	See definition in regulation 2(1) and *Requirements for the approval of dosimetry services* Part 3[43]
Effective dose	See regulation 2(6) and paragraphs 185-188
Employment medical adviser	See definition in regulation 2(1) and paragraph 445
Engineering controls and design features	See paragraph 76
Equivalent dose	See regulation 2(6) and paragraphs 185-188
External radiation	Generally refers to ionising radiation coming from outside the body of an individual, notably X-rays, gamma-rays, beta particles or neutrons (see definition in regulation 2(1)
Health record	See paragraph 458 and Schedule 7
HSW Act	Health and Safety at Work etc Act 1974
Internal radiation	Means ionising radiation coming from inside the body of an individual from inhalation or ingestion or incorporation of radioactive material

171

Local rules	See paragraphs 272-276
Medical exposure	See definition in regulation 2(1) and paragraph 527
MHSWR	Management of Health and Safety at Work Regulations 1999
Naturally occurring radioactive materials	See paragraphs 12 and 13
Outside worker	See definition in regulation 2(1) and paragraph 1
Practice	See definition in regulation 2(1)
Personal protective equipment (PPE)	See paragraph 115
Qualified person	See paragraph 357
Radiation accident	An accident where immediate action would be required to prevent or reduce the exposure to ionising radiation of employees or any other people
Radiation employer	See definition in regulation 2(1) and guidance to Table 1 in the Introduction
Radiation passbook	See definition in regulation 2(1), paragraph 398 and Schedule 6
Radiation protection adviser (RPA)	See definition in regulation 2(1) and paragraphs 4-8
Radiation protection supervisor (RPS)	See paragraph 290-292
Radioactive substance	See definition in regulation 2(1) and associated guidance in paragraphs 9-13
RSA 93	Radioactive Substances Act 1993
Safety features	See paragraph 77
Safety representative	Refer to the Safety Representatives and Safety Committees Regulations 1977
Sealed source	See definition in regulation 2(1) and paragraph 479
Site radiography	See paragraph 249
Supervised area	See definition in regulation 2(1) and guidance to regulation 16(3)
Warning device	See paragraph 78
Woman of reproductive capacity	See definition in regulation 2(1) and paragraph 194
Work with ionising radiation	See definition in regulation 2(1) and paragraph 14

Appendix 3   References

1   *The Ionising Radiations Regulations 1999* SI 1999/3232 Stationery Office 1999 ISBN 0 11 085614 7

2   *Health and Safety at Work etc Act 1974* HMSO 1974 ISBN 0 10 543774 3

3   'Council Directive 96/29 Euratom of 13 May 1996 laying down basic safety standards for the protection of health of workers and the general public against the dangers arising from ionising radiation' *Official Journal of the European Communities 1996* **39** (L159) 1-114

4   'Council Directive 97/43/Euratom of 30 June 1997 on health protection of individuals against the dangers of ionising radiation in relation to medical exposure' *Official Journal of the European Communities* 1997 **40** (L180)

5   *Nuclear Installations Act 1965* HMSO 1965 ISBN 0 10 850216 3

6   *Radioactive Substances Act 1993* HMSO 1993 ISBN 0 10 541293 7

7   *Management of Health and Safety at Work Regulations 1999* SI 1999/3242 Stationery Office 1999 ISBN 0 11 085625 2

8   *HSE statement setting out criteria of competence for individuals and bodies intending to give advice as radiation protection advisors:* available from HSE Radiation Protection Policy, Rose Court, 2 Southwark Bridge, London SE1 9HS

9   *Prior authorisation under the Ionising Radiations Regulations 1999* IRIS5 HSE Books (due to be published summer 2000)

10   *The Safety Representatives and Safety Committees Regulations 1977* SI 1977/500 HMSO 1977 ISBN 0 11 070500 9

11   *A guide to the Health and Safety (Consultation with Employees) Regulations 1996. Guidance on Regulations* L95 HSE Books 1996 ISBN 0 7176 1234 1

12   *The Ionising Radiation (Protection of Persons Undergoing Medical Examination or Treatment) Regulations 1988* SI 1988/778 HMSO 1988 ISBN 0 11 086778 5

13   *Safety signs and signals. The Health and Safety (Safety Signs and Signals) Regulations 1996. Guidance on Regulations* L64 HSE Books 1996 ISBN 0 7176 0870 0

14   *Personal protective equipment at work. Personal Protective Equipment at Work Regulations 1992. Guidance on Regulations* L25 HSE Books 1992 ISBN 0 7176 0415 2

15   *The selection, use and maintenance of respiratory protective equipment: A practical guide* HSG53 HSE Books 1998 ISBN 0 7176 1537 5

16   *Safety assessment principles for nuclear plants* HMSO 1992 ISBN 0 11 882043 5

17   *Employment Protection (Consolidation) Act 1978* Ch44 1978 ISBN 0 10 544478 2

18   *Employment Rights Act 1996* HMSO 1996 ISBN 0 10 541896 X

19   *New and expectant mothers at work: A guide for employers* HSG122
HSE Books 1994 ISBN 0 7176 0826 3

20   *Suspension from work on medical and maternity grounds* PL705 DTI
(available from DTI)

21   *Maternity rights: a guide for employers and employees* DTI/HSE 1995
PL958 DTI (available from DTI)

22   Personal Protective Equipment (Design, Manufacture, Marketing)
Directive 89/686/EEC 'Council Directive of 21 December 1989 on the
approximation of the laws of the member states relating to personal protective
equipment' *Official Journal of the European Communities 1989* **32** (L399) 13-38

23   *Personal Protective Equipment (EC Directive) Regulations 1992*
SI 1992/3139 HMSO ISBN 0 11 025252 7

24   'Council Directive of 30 November 1989 on the minimum health and
safety requirements for the use by workers of personal protective equipment at
the workplace (third individual directive within the meaning of Article 16(1) of
Directive 89/931/EEC) (89/656/EEC)' *Official Journal of the European
Communities 1989* **32** (L393) 18-28

25   British Institute of Radiology 'Patients leaving hospital after
administration of radioactive substances' *British Journal of Radiology* **72** 1999
121-125

26   *Workplace health, safety and welfare. Workplace (Health, Safety and
Welfare) Regulations 1992. Approved Code of Practice and guidance* L24 HSE
Books 1992 ISBN 0 7176 0413 6

27   National Physical Laboratory *Measurement good practice guide: The
examination, testing and calibration of portable radiation protection instruments.*
Available from the National Physical Laboratory, Teddington, Middlesex,
TW11 0LW

28   *Radiation doses - assessment and recording* IRIS2(rev) HSE Books 2000

29   *Data Protection Act 1998* Ch 29 The Stationery Office 1998
ISBN 0 10 542998 8

30   *Central index of dose information* (reports available on HSE's home page
http://www.hse.gov.uk)

31   *Special entries in dose records* (HSE information document available on
HSE's Health Directorate website www.hse.gov.uk/hthdir/specent.htm)

32   *Radiation protection - sealed radioactive sources - leakage test methods*
ISO 9978: 1992

33   *Atomic Energy and Radioactive Substances: Radioactive Material (Road
Transport) (Great Britain) Regulations 1996* HMSO 1996 ISBN 0 11 054742 X

34   *Packaging, Labelling and Carriage of Radioactive Material by Rail
Regulations 1996* SI 1996/2090 HMSO 1996 ISBN 0 11 062921 3

35   *Reporting of Injuries, Diseases and Dangerous Occurrences Regulations 1995*
SI 1995/3163 HMSO 1995 ISBN 0 11 053751 3

36   *Medical Devices Regulations 1994* SI 1994/3017 HMSO
ISBN 0 11 043255 X

37   Medical Devices Directive 93/42/EEC 'Opinion on the proposal for a
council directive concerning medical devices' *Official Journal of the European
Communities 1992* 5 NoC79 1-3

38   *Safe use of work equipment. Provision and Use of Work Equipment
Regulations 1998. Approved Code of Practice and guidance* L22 HSE Books 1998
ISBN 0 7176 1626 6

39   *Electricity at Work Regulations 1989* SI 1989/635 HMSO 1989
ISBN 0 11 096635 X

40   *Fitness of equipment used for medical exposure to ionising radiation* PM77
HSE Books 1998 ISBN 0 7176 1482 4

41   *Requirements for the approval of dosimetry services under the Ionising
Radiation Regulations 1999 Part 1: External Radiations* HSE 1999 RADS
1/1999. Available from the Dosimetry Services Administrator, Technology
Division, Magdalen House, Stanley Precinct, Bootle, Merseyside L20 3QZ

42   *Requirements for the approval of dosimetry services under the Ionising
Radiation Regulations 1999 Part 2: Internal radiations* HSE 1999 RADS 2/1999.
(For availability, see reference 41.)

43   *Requirements for the approval of dosimetry services under the Ionising
Radiation Regulations 1999 Part 3: Co-ordination and Record-keeping* HSE 1999
RADS 3/1999. (For availability, see reference 41.)

While every effort has been made to ensure the accuracy of the references
listed in this publication, their future availability cannot be guaranteed.

HSE Books

Copies of HSE Books publications are available from the address on the inside
back cover.

Stationery Office

Copies of Acts and Regulations are available from the Stationery Office at the
following address:

The Publications Centre, PO Box 276, London SW8 5DT
Telephone orders 0870 600 5522; Fax orders 0870 600 5533

Printed and Published by the Health and Safety Executive   C45   3/00